Design Chronicles

Significant Mass-Produced Designs of the 20th Century

Carroll M. Gantz, FIDSA

edited by Victoria Matranga, H/IDSA

Schiffer Publishing Ltd®

4880 Lower Valley Road, Atglen, PA 19310 USA

Dedicated to my wife, Lorraine

Published by Schiffer Publishing Ltd.
4880 Lower Valley Road
Atglen, PA 19310
Phone: (610) 593-1777; Fax: (610) 593-2002
E-mail: Info@schifferbooks.com

For the largest selection of fine reference books on this and related subjects, please visit our web site at **www.schifferbooks.com**
We are always looking for people to write books on new and related subjects. If you have an idea for a book please contact us at the above address.

This book may be purchased from the publisher.
Include $3.95 for shipping.
Please try your bookstore first.
You may write for a free catalog.

In Europe, Schiffer books are distributed by
Bushwood Books
6 Marksbury Ave.
Kew Gardens
Surrey TW9 4JF England
Phone: 44 (0) 20 8392-8585; Fax: 44 (0) 20 8392-9876
E-mail: info@bushwoodbooks.co.uk
Free postage in the U.K., Europe; air mail at cost.

Library of Congress Cataloging-in-Publication Data

Gantz, Carroll M.
Design chronicles : significant mass-produced designs of the 20th century / Carroll M. Gantz.
p. cm.
ISBN 0-7643-2223-0 (hardcover)
1. Design—History—20th century. 2. Design, Industrial—History—20th century. I. Title.
NK1390.G36 2005
745.2'09'04075—dc22

2005009046

Type set in Helvetical heading font/text font Helvetica

ISBN: 0-7643-2223-0
Printed in China

Contents

Introduction

Daily, we use and observe hundreds of utilitarian objects so common that we scarcely notice them, let alone consider their origins. Archeologists reconstruct past cultures by excavating antiquity's everyday objects. Thus, pottery, weapons, and tools provide essential information about people long gone. In like manner, today's man-made artifacts reveal our unique culture and values. Each item was carefully conceived and designed by individuals who considered users, function, materials, appearance, and configuration. This creative process, this combination of cultural values and technology, is known today as industrial design, because this process results in artifacts that are produced in large quantities by industry, not by the original designer or a craftsman.

The 20th century encompasses the emergence, development, and refinement of what we often refer to as modern design—a visual style characterized by functionality, minimal ornament, and simplicity of form. Many design historians judge excellence by the degree to which artifacts meet the arbitrary visual criteria established by museums for such movements as Art Nouveau, Art Deco, International Style, etc., just as art history has been categorized into periods of visual commonality.

But this style-oriented categorization offers only a cosmetic appreciation of 20th century design. What distinguishes this 20th century design from all earlier art or design is its deliberate merger with the uniquely American idea of mass production and consumption. For the first time in history, artifacts were produced in enormous quantities, and the designer was required to design for new contexts of mass manufacture, mass merchandising, mass distribution, and mass consumption. This environment produced prosperity, democratized wealth and led to the nation's superpower status—the very essence of the American Century. The fundamental legacy of the 20th century is the emergence of rapid technological change through U.S. economic strength. Industrial design, the process that converts technology into desirable, appropriate, and needed material goods for mass consumption, has been a vital participant in this miracle.

This book celebrates this unique evolution. American design by 1920 had embraced a new corporate-based research and development system, a market-based economy, and a mass-distribution system, which, by 1960, allowed the middle-class to acquire consumer goods beyond the wildest dreams of the 19th century. The design of these goods was significant, not merely because of style, but because the designs held economic substance that contributed greatly to the steady growth of the U.S. Gross National Product, by 2004 over $10.5 trillion; roughly equal to that of Japan, Germany, Great Britain, France, and Italy combined.

There are many books on design history, but this is the first to arrange design information into a year-by-year chronology. Why? Because this is the way design thinking actually developed—in a sequential manner, and within a contemporary context of technology, culture, people, and current events. A chronology reflects this reality. In a similar way, each year is a cross-section of its own time, technologies, materials, ideas, and culture. Finally, a chronology, unlike history presented as an orderly lineage of movements or styles, identifies the actual sequence of who did what, when, where, and why.

Each year of the 20th century appears as a double-page spread that cites and illustrates significant product designs and events of that year. Product design is highlighted, not only because of its cultural influence, but also because of the tremendous economic and functional impact of such mass-produced artifacts upon the daily life and well being of millions. Wherever possible, designers are named to remind us all that individual creativity, intellect, and skill are the sources of successful products. Cross-references to other years allow the reader to follow a particular event, company, or product across decades. The primary question in a chronological arrangement is how to place products into a specific, appropriate single year. Generally, this book lists products and events in the first year in which they were introduced in quantity or became public.

Possible product choices are so numerous that selection is painfully arbitrary. Many readers may find some of their favorites omitted or condensed in detail. While this is regrettable, it is my hope that such shortcomings are compensated for by easier reference, the variety of information, and, most importantly, the condensation into convenient year-at-a-glance pages. Some are "classics" of "good" design selected by arbiters of taste or design awards. Others are notorious failures in the marketplace, reminding us that public acceptance has its limits. Some are technological or merchandising "firsts," or those that set a "typeform" (a commonly accepted "correct" form) for a specific product. Finally, some simply have been very popular and successful. In a sense, approval by the public indicates that a product satisfied the human needs, desires, and values typical of the time.

The primary purpose of this book is to be a general reference for designers, students, collectors, and laypersons. It is my hope that this chronology provides a useful guide to design as well as a tribute to all designers, manufacturers, and institutions, contributing to its fundamental purpose: that of creating and optimizing the utility, value, and appearance of our man-made environment for the largest number of people.

Acknowledgments

When I began this book, I naïvely thought of it as an independent project. I soon learned that any book is dependent upon many individuals and organizations for its development and publication. It is those who provided me graciously with their patience, support, assistance, and information that I now gratefully acknowledge.

The Industrial Designers Society of America (IDSA) has been particularly central to the concept of this book. The Society in 1993 offered to publish a series of my articles titled *Design Time Capsules* in IDSA's professional journal, *Innovation*. Executive Director Kristina Goodrich encouraged the idea of a similar chronological book and provided access to illustrative materials from IDSA files and Industrial Design Excellence Awards (IDEA) files (starting 1980) which were invaluable to this book from that date forward.

The archival records of the Industrial Designers Institute (IDI), a predecessor organization to IDSA, reside in a special collection of the George Arents Library at Syracuse University, established by the late Professor Emeritus Arthur J. Pulos, FIDSA, a respected source of inspiration for this book. Terry Keenan and Carolyn Davis of the library were particularly helpful. Many of the product illustrations from 1951-1965 are from the IDI awards program of this period. Many outstanding art, design or public museums across the country provided research information and illustration resources. All are appropriately credited.

I would like to especially recognize the steadfast personal inspiration and professional contributions of Victoria Matranga, Design Programs Coordinator of the International Housewares Association (IHA). Over the years she has shared her considerable knowledge and information of design history with me, and her careful reading and editing of this manuscript improved the text. She also lent several IHA photographs as well as her own.

Invaluable indeed were many fellow IDSA members and friends, most of who are well known in the industrial design profession because of their own work and contributions to the design community. Their recollections, shared information, illustrations, and their personal support of this book will be long remembered. I am indebted to many IDSA colleagues who shared personal experiences and specialized design information, including Reed Agnew, Andrew L. Alger, Charles L. Allen, Wallace H. Appel, Curtis Bailey, James G. Balmer, Arthur N. BecVar, Denis Bedel, Robert Blaich, Andrew Cogan, Alvaro Correa, Michael Cousins, Morison Cousins, Tim Cunningham, Donald E. Dailey, Thomas M. David, Vincent Foote, Robert Gruen, Marc Harrison, Jon W. Hauser, Sam Highberger, Samuel E. Hohulin, Marshall B. Johnson, Robert Kalvitis, Henry Keck, Douglas Kelley, Rudy Krolopp, Al Lambach, Michael Lax, James J. Lesko, Monte Levin, Tucker P. Madawick, Ellen Manderfield, Tom Matano, Douglas J. Medema, Lee Payne, Mark Penke, James Pirkl, Gerald E. Proctor, Alan G. Reddig, William E. Reddig, James M. Ryan, Clair Samhammer, Roger Schneeman, Kenneth C. Schory, Bret H. Smith, Robert G. Smith, Paul B. Specht, Raymond Spilman, Budd Steinhilber, Keiichi Totsuka, Arnold Wolf, Douglas F. Wolff, Cooper Woodring, Jerry E. Wright, Donald W. Zurwelle, and many others. Without their generous assistance and support, this book would not have been possible.

I am grateful also for special information from Anita Bach, William Blitzer of Lightolier, Clive Chajet of Lippincott & Margulies, Philip S. Egan, Barton Friend, Sharon Hite, Ron Hodges of Steelcase, Inc., John Jacobus, Tom Kellogg, Allen Koenigsberg, Mary Ann Latham, J. Gordon Lippincott, Michael McCue of Kangaroo Products, Eudorah Moore, Chip Nowacek, John Okolowicz, Joseph J. Pankus, Gordon Peacock, Barbra Reinecke Sedory, Enrico Tedeschi, and Paul Wiedmann of Peter Muller-Munk Associates.

While many manufacturers provided the information and illustrative materials I requested, I would like to mention several that demonstrated a deep appreciation of its own place in design history as well as gracious and unusually generous support through its representatives. The Hall of History, a privately supported archive of the General Electric Company, through Mary Kuykendall, its Administrative Manager, and John Anderson, provided generous research time and informational support. Herman Miller, Inc., through corporate archivists Robert Viol and Richard Rutledge, provided superlative service, information, and illustrations. The Knoll Group, through Communications Manager Messeret Sertse, was also extremely generous in its illustrative materials. Eastman Kodak, through Senior Designer Alan G. Reddig, IDSA, generously provided many special photographs of the company's most famous products throughout the century. The credits for photographs identify the many corporate sources that provided invaluable assistance.

Prologue: 1800-1899

Frank Lloyd Wright was seven years old when Gen. George Custer was killed at the Battle of Little Big Horn. Thus the life of one of the most prominent design figures of the twentieth century connected to the frontier of American history. Until 1900, the United States was too preoccupied domestically to consider the meaning of art and design, let alone influence its course. This was left to Europeans, considered the source of all artistic innovation. Classicism, the picturesque imitation of historical styles and Romanticism, the embellishment of products with literal visual icons recalling emotional qualities of spirituality, nature, or virtue, drove European art and design. Ordinary household products for the wealthy were conceived as works of fine art and handcrafted by skilled artisans.

The late 18th century British Industrial Revolution, based on steam power, mechanized the fabrication of products. The U.S. pioneered the use of interchangeable parts in production, primarily in firearms, in the 1830s. American Shaker design (1825-1850), the earliest evidence of simple, functional design, was inspired neither by style nor mass-production, but by spiritual values that required the making of only useful articles. Shaker designs were spurned by Americans and ridiculed by the English as old-fashioned and totally lacking in style. Traditional design dominated the 1851 "Great Exposition" at the Crystal Palace in London, conceived by Henry Cole, a British civil servant concerned about the lack of design and production standards in British industry. Some criticized American products at the Crystal Palace (such as the Colt revolver, McCormick reaper, and John Deere's steel plow) as tasteless and severe because they lacked ornament. But others recognized such products were developed to serve a specific human need and to reach a mass market.

Soon, modern technology began to shape the 19th century. In Austria, Michael Thonet developed new processes for furniture manufacture in 1857. New technologies included the telegraph (1837), the typewriter (1873), dry-plate photography (1873), roll film (1884), popular consumer cameras (1888), motion pictures (1895), and the cathode ray tube (1897). Experimenting with electricity in 1878, Thomas A. Edison and Sir Joseph Swann, working independently, produced the first incandescent lamps. In 1882 Edison's generating station initiated economical power for home and industry. In 1889, Nikola Tesla patented the multiphase electric motor, and, in 1891, George Westinghouse produced a small, portable electric fan. These innovations and others, such as the telephone (1876), phonograph (1877), automobile (1886), radio (1895), electric railways (1896), and recording systems (1898), were individual components of what would grow into vast technological systems of telecommunications, photography, entertainment, electrification, and transportation. This became known as the Second, or American, Industrial Revolution, based on electric power, machine tools, and factories, which developed the "American System of Manufacture."

Some in England viewed the Industrial Revolution as a threat to civilization. Monotonous labor tasks, despicable working conditions, and mechanical shapes and forms were offensive to artists and intellectuals who feared the loss of human dignity. In the late 1800s the English Arts and Crafts Movement, led by John Ruskin, William Morris, and C.R. Ashbee, responded to such concerns by promoting ideals of craftsmanship. This movement influenced many designers in Europe and in the U.S.

Art Nouveau, inspired by Celtic organic patterns and symbols, took its name from a Parisian shop of that name opened in 1895 by the German art dealer Siegfried Bing. He sought out the most daring craftsmen such as Louis Tiffany in the U.S. and Henry Van de Velde of Belgium, the country Bing credited with originating the style. *Art Nouveau* swept Europe as *Le Style Moderne*, or *Le Style Anglais* (France) *Neue Stil* or *Jugendstil* (Germany), *Modernisme* (Spain), and *Lo Stile Liberty* (Italy). The term "Liberty" came from the English merchandiser, Liberty & Company, for whom designers Christopher Dresser (1834-1904) and Archibald Knox (1864-1933) worked.

In 1896, Elbert Hubbard, inspired by the Arts and Crafts Movement, founded a workshop, *Roycroft*, in East Aurora, New York. In 1897, Peter Behrens, Richard Riemerschmid (1868-1957), and others formed the Munich *Vereinigte Werkstätten für Kunst im Handwerk* (United Workshops for Art in Handcraft), while Josef Hoffmann, Otto Wagner, and others founded the *Wiener Sezession* (Vienna Secession), the Austrian version of *Art Nouveau.* In 1898, Karl Schmidt in Germany founded the *Dresdener Werkstätten für Hand-werkskunst* (Dresden Workshops for Handicraft Arts), and, in the U.S., Gustav Stickley, formed "Craftsman" to produce furniture known as Mission Style. In 1899, the *Künstler-Kolonie* (artist's colony) was founded in Darmstadt, Germany.

The 1893 World's Colombian Exposition in Chicago introduced the general public and designers to the international design scene. The Fair also dramatized the impact of technology and industrialization on daily life with exhibits such as a "Model Electric Kitchen" which included an electric range, broiler, and water heater. The Spanish-American War, in 1898, and the end of the 19th century awakened America to its growing international role in trade, technology, industry, manufacturing, architecture, culture, and politics. By the end of the century, the country was ready technologically and psychologically to become a global leader in mass manufacture and distribution of consumer products, in the development of new technologies, and in the exploration of new design directions.

"Mass" Production

As emphasized in the introduction, the economic phenomena of mass production, mass merchandising, mass distribution, and mass consumption established the U.S. as a global leader, and contributed to the popular view of the 20th as the "American" Century.

This phenomenon of mass consumerism became the defining principle of industrial design as it developed and flourished. Its founders embraced the new challenges with enthusiasm, many of them coming from disciplines already attuned to popular culture, such as advertising or theater. These designers willingly accepted artistic, manufacturing, and cost restraints in order to provide a broad public with customer features they desired (comfort, safety, functionality, affordability), and to provide manufacturers with profits.

One characteristic that distinguishes "industrial" design from other design is the fact that mass manufacture and distribution result in acceptance or rejection by a large user base, rather than by an exclusive patron or elite segments of the population. Such design demands special sensitivity to fundamental and realistic human needs—physical, aesthetic, economic, and emotional; as well as understanding the business of capital investment, retail competition, consumer research, and profitability. Industrial design is "art" with a purpose. That purpose is market success.

Understandably, museums and art historians have been less concerned with market success than with visual characteristics that represented current design fashions or established aesthetic standards. The Museum of Modern Art, from the 1940s to the 1970s, determined "good design," primarily on the basis of what was known as the "International Style," now a relic of design history, partly because it was based on style rather than economic substance or popular acceptance.

But I submit that mass-market acceptance of a particular design over a long period of time is a valid criterion for "good" design as well. Market acceptance means that a design survived the fleeting accolades of current fashion, design awards, museum approval, or personal artistic statements, and won long-term approval by its intended users while providing economic benefit to investors and manufacturers. In fact, many of the century's icons of design; the Model T, the Coke bottle, the Jeep, blue jeans, the KitchenAid mixer, Volkswagen Beetle, and Zippo lighter, among others, were never recognized by museums in their time. But they had lasting economic and cultural substance.

One way to measure these criteria of mass acceptance is to look at sales volume (economic substance) over a period of time. Unfortunately but understandably, manufacturers are often reluctant to share such competitive data. But for a sense of comparative scale for some of the familiar products mentioned in this book, a few examples for which data are available follow:

Quantity (in millions)	Item(s)	Time	See year
100,000	Crayola Crayons by Binney & Smith	1903-1998	1903
2,000	*Corelle* LivingWare by Corning (pieces)	1970-1996	1970
1,000	*Barbie* Doll by Mattel	1959-1998	1959
500	*Fiesta ware* by Homer Laughlin (pieces)	1936-1997	1936
400	*Zippo* cigarette lighter	1933-1998	1933
250	*Slinky*	1945-1995	1945
186	Sony *Walkman*	1979-1999	1979
150	*Dustbuster* by Black & Decker	1978-2004	1978
100	VW original *Beetle*	1938-1999	1938
100	Gillette *Sensor for Women* razor	1992-1999	1992
93	Model *500* Bell desk telephone	1949-1982	1949
80	*American Modern* dinnerware (pieces)	1939-1959	1939
50	Kodak Brownie camera	1900-1910	1900
50	Thonet *Vienna Cafe* Chair #14	1876-1900	1900
50	Revere *Copper-Clad Steel Cookware*	1939-1957	1939
50	Autos designed under Harley Earl	1928-1959	Various
20	Toyota *Corolla*	1963-1996	1959
20	Plastic Pink Flamingos	1957-1996	1957
16	Ford Model T	1908-1927	1908

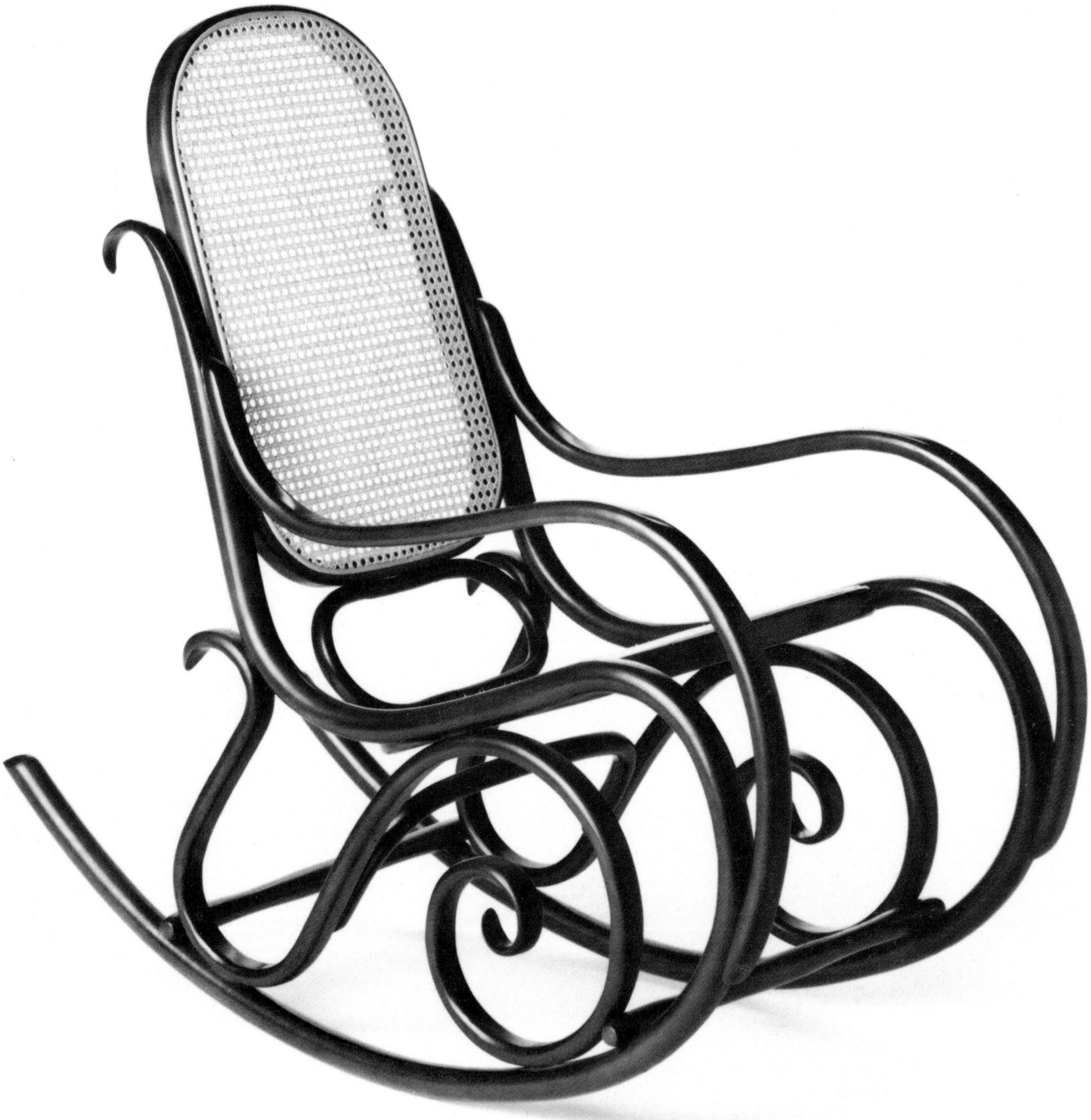

Chapter 1

1900-1909 Affordable Design

The end of the 19th century has a dramatic effect on the U.S. psyche. The Spanish-American War, the assassination of President McKinley, and the Panama Canal Treaty awaken U.S. public awareness of the country's new international leadership role, and its freedom from traditional dependence on Europe. Americans are now ready, not only to buy, but to compete in international markets. Initially American production competes, not on the basis of style, but in product utility, systems for interchangeable parts, and mass production and distribution.

The U.S. demonstrates technological and manufacturing prowess in the making of automobiles, aircraft, and new products such as telephones, inexpensive cameras, phonographs, and electric household appliances. By 1908 Ford's Model T automobile is an international symbol of affordable technology for the masses, and a huge automotive consumer market emerges. New mechanized farm equipment and transportation systems improve farm profitability, and increase disposable income for household improvements and now-attainable luxury items such as decorative ceramics, lamps, and furniture.

Consumer goods such as carriages, furniture, household equipment, and decorative arts had been available in limited quantities and at prices that only the wealthy could afford. Now, the new American system of manufacture provides similar products at lower cost and in higher quantities. Many of these products are labor saving devices such as sewing machines, typewriters, and farm equipment. Since 1896, Rural Free Delivery has brought mail directly to more than 20 million homes, including catalogues from Montgomery Ward and Sears, Roebuck & Co., encouraging a mass consumer market for household products. Though most of the country's population lives in rural areas, by 1909 more than 500 new five-and-ten cent stores, such as F. W. Woolworth's, offer inexpensive household goods and clothing to every small town.

In 1904 the Louisiana Purchase Exhibition in St. Louis, the first major international fair in the U.S., fuels consumer demand. People can see first-hand what Europe and the Orient offer in style, innovation, and exoticism. European Art Nouveau and Arts and Crafts trends are well represented, as are versions of these styles produced by U.S. firms such as Tiffany Studios and Roycroft. These new styles inspire U.S. manufacturers and designers, and the public responds enthusiastically.

General Electric Company and other corporations establish research laboratories that propel many new discoveries. New 20th century materials such as aluminum allow new products such as lightweight kitchen utensils, and the development of plastics begins a century of synthetic materials that will replace the traditional 19th century materials of wood, iron, leather, cotton, and wool.

Scientific research also enhances the everyday life. Through the new science of home economics, consumers become aware that their own kitchens can be more efficient, convenient, and useful. New appliances such as stoves, toasters, and washing machines become available to upper-income households with electrical service. Kitchens will lead consumers, and their homes, into the 20th century.

A vigorous national awakening to international trade competition, and the establishment of a powerful infrastructure of research, development, manufacturing, distribution, and consumption, opens the American century.

01-01

01-02

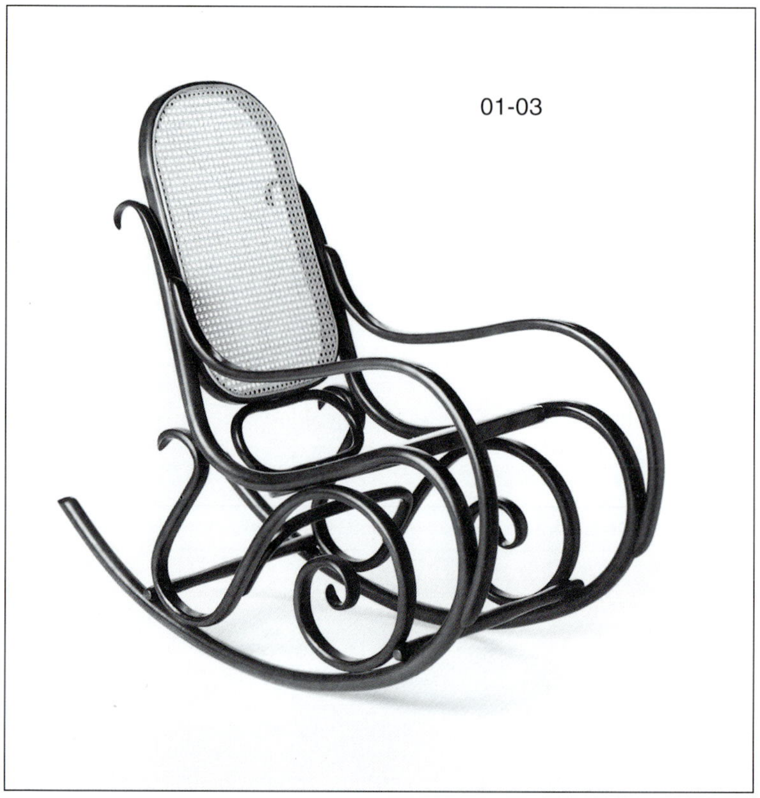

01-03

01-04

01-05

1900

The *Vienna Secession* (see Prologue) exhibits work of C. R. Ashbee (1863-1942) and Charles Rennie Mackintosh at its eighth exhibition devoted to handicraft. Co-founder Josef Hoffmann says the exhibition "strives to achieve an identical goal within any nation to impart appropriate form to modern perception."

Exposition Universelle et Internationale in Paris, the major event of architecture and design this year, establishes *Art Nouveau* as a modern style (see Prologue). The exposition includes the work of French glass and ceramic artist Emile Gallé (1846-1904), founder of the School of Nancy, René Lalique (see 1904, 1910), and Louis Comfort Tiffany. Kodak's Brownie camera and Rookwood pottery are among 7,000 U.S. products shown.

Construction begins for a *Künstler-Kolonie* (artist's colony), founded last year on the Matildenhöhe in Darmstadt, Germany (see Prologue). Josef Maria Olbrich (1867-1908), is designing the buildings for the colony. Construction will conclude in 1907 with the crowning symbol, the "Wedding Tower" exhibition hall. Peter Behrens arrives as a leading professor there. The colony will remain active until 1914.

Gustav Stickley introduces his new line, called *The New Furniture,* at the Grand Rapids (Michigan) Furniture Show. It will become known as "Mission" style Furniture (see 1901)

Gottlieb Daimler (1834-1900), founder of Daimler-Motoren-Gesellschaft (DMG) in 1890, and Wilhelm Maybach (1846-1929), his technical director, fit the "carburettor" he invented in 1892 to Daimler's car. They name it *Mercedes,* after Daimler's most important distributor's daughter, Mercédès Jellinek.

Gebrüder Thonet (see Prologue) has 52 factories in Europe making bentwood furniture. German cabinetmaker, Michael Thonet (1796-1871), and his five sons founded the company in 1853. Its most successful model is the Vienna Café chair No. 14 (see illustration). Other well-known models include the Rocking Chair # 10 (see illustration) and a sensuous reclining couch Model No. 2, produced since 1885. All were designed by Michael Thonet.

01-01. The Kodak *Brownie* box camera is introduced by the Eastman Dry Plate Company, founded by George Eastman (1854-1932). The *Brownie* costs only $1.00, and will popularize photography. Frank A. Brownell designed it to be suitable for children, so its package is adorned with Canadian author Palmer Cox's "brownie" elves, a current cartoon character. By 1910, fifty million *Brownies* will be sold.

01-02. Vienna Café chair No. 14 by Gebrüder Thonet. Produced since 1859. By 1930, 50 million will be produced.

01-03. Rocking Chair No. 10 by Gebrüder Thonet. Produced since 1866

01-04. There are 2000 nickelodeons in Brooklyn alone. The term combines "nickel" and "melody." These coin-operated phonographs, invented in 1889 by Louis Glass, use ear-tubes that look like musical hookah pipes. They will later be called "Jukeboxes." See 1927, 1934, and 1946.

01-05. Architect Frank Lloyd Wright is in the process of completing his studio in Oak Park, Illinois, which he started in 1898. The studio is attached to his house, completed in 1890. The house was totally renovated in 1895, and its new interior and furniture, including these dining room chairs, is the first evidence of his new, spare, uncompromising style.

01-06. AT&T, created in 1896 to provide long-distance telephone service, acquires American Bell Telephone Company, founded in 1877 by Alexander Graham Bell (1847-1922) and his father-in-law Gardner Hubbard. It becomes the sole provider of telephone service in the U.S. There are now one million telephones in use, like this "Common Battery" 1896 wall phone, whose electric power source is a central exchange.

01-06

01-07

01-08

01-09

1901

Peter Behrens designs his own home, the only home at the *Künstler-Kolonie* in Darmstadt (see 1900) not designed by Austrian architect Josef Maria Olbrich (1867-1908). Behrens, a professor at the colony, designs everything in the house including cutlery, door handles, and furniture.

The achievements of the Darmstadt *Künstler-Kolonie* (see 1900) are exhibited in *Ein Dokument Deutscher Kunst* (A Document of German Art). Peter Behrens organizes the event and designs its official poster. 10,000 attend.

Gustav Stickley changes the name of his company in Syracuse, New York, to "Craftsman," adapts the motto *"Als Ich Kan* ("If I can"), and exhibits what will become known as "Mission" style furniture. The term refers loosely to the work of various builders influenced by the Stickley family, whose "mission" is simplicity, durability, and quality. Gustav Stickley begins to publish his magazine, *The Craftsman*, which will continue until 1915.

The Toledo, Ohio Museum of Art is established by Edward Drummond Libbey, who had opened a glass factory in Toledo in 1888. The first prominent director of the Museum is George W. Stevens. In 1918, the Museum will establish a School of Design that will survive until 1942. See 1928.

The first "mechanical" building game is invented and patented by English game designer Frank Hornby (1863-1936). It consists of compatible metal pieces meant to stimulate a child's creative spirit and a taste for mechanics. It will be marketed in 1906 as *Mechanics Made Easy,* but will later be known as *Meccano.* Hornby will also produce electric model train systems bearing his name in the 1920s, and a range of die-cast model vehicles known as "Dinky" (Scottish slang meaning small and neat) toys in the 1930s.

01-07. English artist Francis Barraud painted the painting that became a trademark, "His Master's Voice," in 1898-1899, to show the reaction of his brother's dog, *Nipper*, to phonograph music. The painting was sold to Emile Berliner's Gramophone Company, Ltd., in London in 1899, and American rights were sold to the Victor Talking machine Company this year for use on its product. The trademark will be acquired by RCA in 1929 and used extensively until discontinued in 1968. But its public appeal will have become so pervasive and famous that it will be rejuvenated in 1990.

01-08. The Victor Talking Machine Company, formerly Berliner's U.S. Gramophone Company, introduces a new hand-cranked *Victor Talking Machine.* It was patented and produced by Emile Berliner (1851-1929) in 1895 as the *Gramophone.* Berliner and Camden, New Jersey, inventor Eldridge R. Johnson, who improved the sound quality of Berliner's phonograph disc (now called a "record" system), form the new company this year. The Victor Company will be purchased by RCA in 1929. See 1906.

01-09. Olds Motor Works (founded 1897 by Ransom Eli Olds, 1864-1950) debuts the first U.S. gasoline-powered auto to be produced in quantity, the *425 Runabout.* It features a new, gracefully curved dashboard. It will be produced until 1907, and the U.S. Postal Service will use some as the first mail trucks. See 1908.

01-10. The Automatic Electric Company introduces a tabletop telephone similar in style but with cleaner lines than Bell's model of 1896. It includes the Strowger finger dial, invented and first produced by Alman Brown Strowger, who founded the Company in LaPorte, Indiana in 1892. Strowger also invented the automatic switchboard number-dialing system in 1889, and initiated it in 1892. The photo shows an 1896 Strowger dial wall phone.

01-10

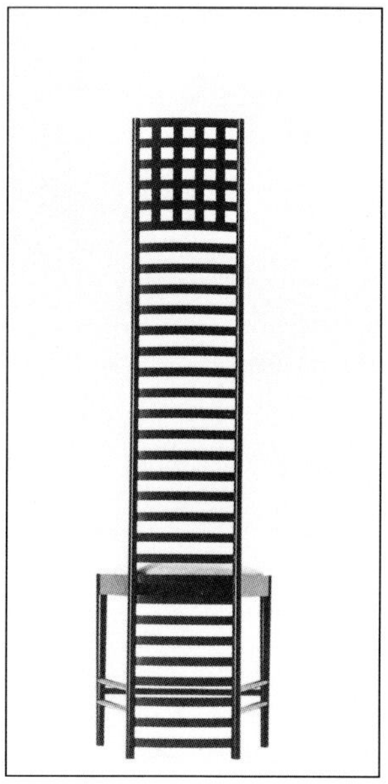

01-11

01-12

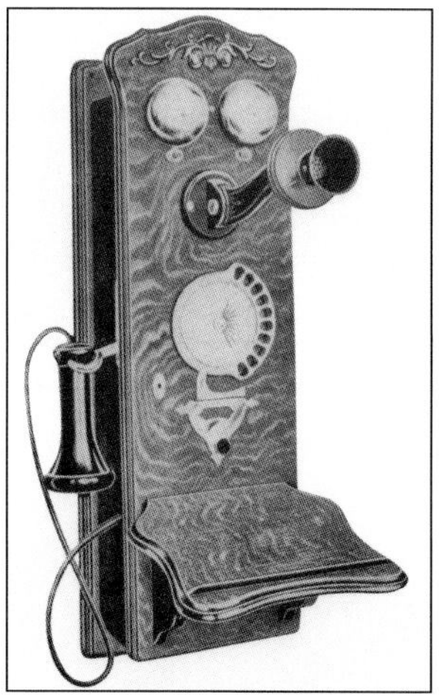

01-13

1902

Painter Aladár Körösfói-Kriesch, Swedish craftsman/designer Leo Belmonte, and architect István Megyaszay found the *Gödölló Workshops* in Hungary. They produce crafts in the Arts and Crafts ideal, and will achieve great success at the Louisiana Purchase exposition in 1904. The workshops will receive Hungarian State support from 1907 until 1921.

Inspired by the Arts and Crafts movement, Henry Van der Velde establishes *Weimar Kunstgewerbeschule* (Weimar School of Arts and Crafts) in Weimar, Germany, to achieve cooperation between artist, craftsman, and industrialists. The school will become public in 1907, and will later become the Bauhaus (see 1919).

The "Teddy Bear" doll is produced by Morris Michtom, based on a political cartoon of "Teddy" Roosevelt refusing to shoot a bear cub. The Steiff Company in Germany, which made earlier "bear" dolls, is deluged with huge orders from the U.S. The Butler Brothers wholesale firm will buy Michtom's entire output of 1903, and will become Ideal Novelty and Toy in 1938.

Louis Vuitton (1821-1892), successful French luggage maker, introduces a line of trunks and cases to the U.S. Designs are by his son Georges (1857-1936), and are distinguished by Louis's initials, adopted in 1898, to discourage imitation.

01-15

Glass blower Reinhold Burger of Germany introduces a device called the *Thermos.* It is based on silvering the interior of a double-walled glass vessel with a vacuum between, developed in 1893 by Scotsman Sir James Dewar, who neglected to patent his invention. It will be patented next year by Burger to form Thermos Gesellschaft M.B.H. in Berlin.

Alexander Kerr designs and introduces his *Economy* jar lid, combining the lid and gasket as a single unit. It is an improvement on the separate lids and gaskets patented in 1858 by John Landis Mason (1832-1902).

The first Cadillac automobile debuts, built by the chief engineer of the new Cadillac Automobile Company, Henry Leland (1843-1932). The first Studebaker (electric car) also appears.

01-11. Charles Rennie Mackintosh designs this high ladderback bedroom chair and a home, known as the Hill House, for publisher Walter W. Blackie in Helensburg, Scotland. Alexander Martin, a local joiner, builds the chair.

01-12. Bernard Eli "Barney" Oldfield (1878-1946) drives Henry Ford's "999" racing car to set a new speed record of 60 mph, a "mile a minute!" In the photo, Henry Ford is standing, and Oldfield is in driver's seat. By 1904 Oldfield will hold most dirt track records from 1 to 50 miles, and by 1906 will become an actor on Broadway.

01-13. The Automatic Electric Company's new wall telephone features more highly decorative details and natural wood grain than last years model, as well as the company's unique design feature, a rotary finger dial (see 1901).

01-15. Otto Wagner, considered by many to be the first "modernist" architect, designs *Die Zeit* telegraph office in Vienna, a classic example of the *Wiener Sezession* (Vienna Secession) founded in 1897 by Wagner and others. (Photo from a 1986 re-creation)

01-16

01-17

01-18

01-19

1903

A perfected bottle-blowing machine, patented by Ohio glassmaker Michael Joseph Owens (b. 1859) in 1895, is now producing light bulbs and bottles at a rate of 2500 per hour. In 1929 the Owens Bottle Company will acquire Illinois Glass Company to become Owens-Illinois Glass Company.

Wear-Ever aluminum pots and pans are demonstrated at Wanamaker's department store in Philadelphia. The public responds enthusiastically to the quick-heating material although the cookware is expensive. Such aluminum cookware will create a new housewares industry.

Josef Hoffmann (see Prologue) and Koloman Moser create The *Wiener Werkstätte* (Vienna Workshops) in Austria with an official seal designed by Hoffmann. An avant-garde group modeled on Charles Robert Ashbee's (1863-1942) London Guild of Handicrafts, the *Werkstätte* advances hand production techniques. Products include bookbinding, metalwork, leatherwork, jewelry, furniture, and architecture. The group's first exhibition will be in 1904 and, by 1905, it will employ over 100 craftsmen. They develop an international reputation until the workshops are liquidated in 1932.

British-American designer Frederick Carder founds Steuben Glass Works in New York. Over the next 30 years he will explore the new techniques and color initiated by Louis Comfort Tiffany. His lustrous *Aurene* technique will produce glass rivaling the *Favrile* designs of Tiffany. The company will be acquired by Corning Glass Works in 1918, but will remain commercially unsuccessful until architect John M. Gates and sculptor Sidney Waugh establish an internal design department in 1936.

William Harley and the Davidson brothers, William, Walter, and Arthur produce the first Harley-Davidson motorcycle in Milwaukee, Wisconsin. The Werner Brothers in France have been manufacturing bicycles with added engines since 1897, but the Harley motorcycle integrates the engine into the frame. See 1936.

01-21

01-16. King Camp Gillette (1855-1932) introduces the safety razor accompanied by 20 disposable blades. After eight years working out practical details with engineer William E. Nickerson, he established the Gillette Safety Razor Company in 1901. Only 51 razors will be sold this year, but next year will see sales of 12.4 million blades.

01-17. Henry Ford (1863-1947) founds the Ford Motor Company and introduces his first Model A car, driven by two cylinders and eight horsepower.

01-18. Bicycle shop owners Orville (1871-1948) and Wilbur (1867-1912) Wright achieve the first powered flight in their *Flyer I*, a plane driven by a 12-hp engine. The first flight in Kitty Hawk, North Carolina, covers 120 feet in 12 seconds. The brothers began their experimental flight program in 1899.

01-19. Binney & Smith introduces *Crayola®* crayons, a patented mixture of paraffin wax and pigment, in a box of eight-colors (red, blue, orange, green, yellow, violet, brown, and black) that sells for a nickel. The name *Crayola* combines the French word "craie," which means chalk, and "ola," short for oleaginous or oily. Edwin Binney and C. Harold Smith founded the company in 1885 to make red oxide paint for barns, carbon black for car tires, and industrial marking crayons. It will be acquired by Hallmark Cards in 1984, and by 1998, will have sold 100 billion crayons.

01-21. Richard Riemerschmid (see Prologue) begins work at the Dresden Workshops (see 1906). His armchair design displays some of the same principles that will later typify the Bauhaus (see 1919).

01-23

01-22

01-24

01-25

1904

The U.S. Commissioner of Labor reports that a stove manufacturer was so concerned about the esthetic qualities of his product that he paid a well-known designer $5000 to design a kitchen stove that would "not offend the eye of the day laborer."

Josef Hoffmann designs one of his signature buildings, the Puckersdorf Sanitorium (1904-1906), a fashionable spa outside Vienna. The project is one of the first commissions for the new *Wiener Werkstätte* (1903); it includes design of all furnishings to be produced by Jacob & Josef Kohn. The first, an armless side chair Model 322, is completed this year for the dining room. In about 1906 a side chair with arms, Model 322/F, will be completed, and in 1908 a matching *Sitzmaschine* (sitting machine) metal reclining chair with adjustable backrest, Model 670. All are memorable Hoffmann styles.

Otto Wagner designs the Imperial Austrian Postal Savings Bank in Vienna (1904-1912) with an innovative and modern glass-and-steel interior. He also creates office furniture to complement the spaces. The furniture, particularly bentwood Chair 247 completed in about 1905, was the first in Austria to use aluminum elements, and was produced for others as well by Gebrüder Thonet (see1900) and Jacob & Joseph Kohn. The furniture will be still in production, slightly altered, in 2000.

Frenchman Louis-Joseph Cartier (1875-1942) devises the first commercially produced wristwatch for dirigible pioneer Alberto Santos-Dumont (1873-1932), a Brazilian living in Paris who built a series of fourteen single seat airplanes, and will achieve the first successful European flight in 1906. Santos had complained he could not effectively use a pocket watch while navigating. The watch is named the "Santos" wristwatch. Men begin to wear wristwatches, formerly considered "feminine."

01-22. Josef Hoffmann produces a silver place setting for himself and other members of the *Wiener Werkstätte* (See 1903). His first cutlery, it will be produced in limited quantities as the *Flaches Modell* (Flat Model) until 1908, and will be first publicly displayed in at a *Werkstätte* exhibition in Vienna (See 1906).

01-23. Frank Lloyd Wright completes the Darwin D. Martin House in Buffalo, New York.

01-24. The Louisiana Purchase Exhibition is held in St. Louis, Missouri, one year late, to celebrate the centennial of the Louisiana Purchase of 1803. Exhibits include the new furniture and interior designs of Peter Behrens, Joseph Maria Olbrich (1867-1908), and Josef Hoffmann, which are seen by Frank Lloyd Wright. Behrens also designs the official catalog for "The Exhibit of German Empire." René Lalique's spectacular Art Nouveau jewelry (see 1910) is an instant success in the U.S. Several new food items are very popular at the fair, including something called a "hamburger," hot-dogs, iced-tea, and ice cream cones.

01-25. Frank Lloyd Wright designs several pieces of furniture, including a curved-back armchair for the Darwin D. Martin House in Buffalo, New York, a swivel office chair, and this straight-backed oak side chair. Called the "Plank Chair," the design will inspire the famous *Red Blue Chair* (see 1918).

01-26

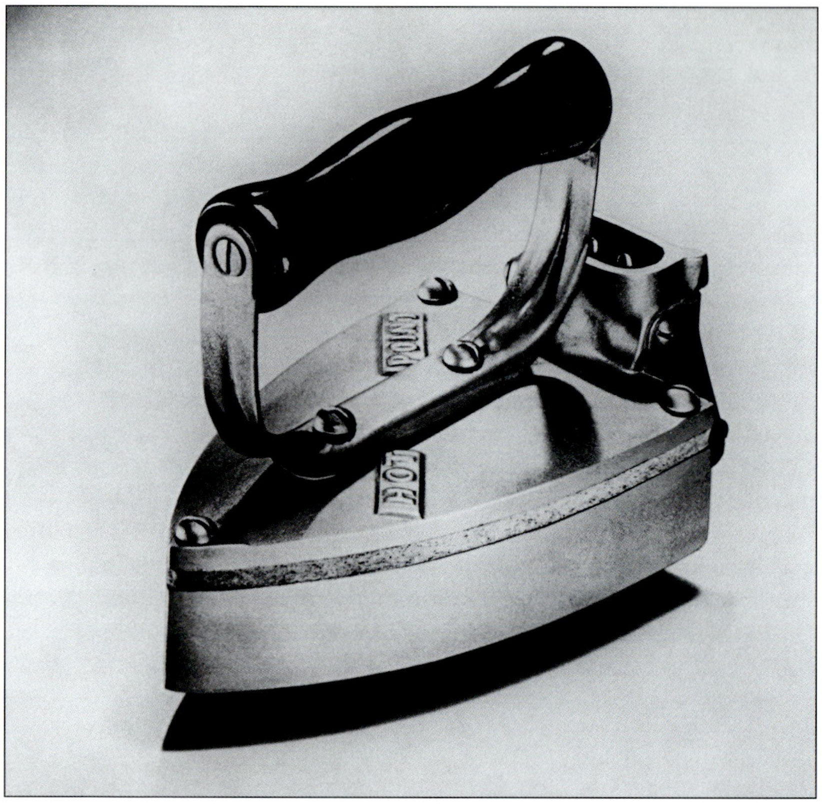

01-27

01-28

1905

Albert H. Munsell (1858-1918), painter and color theorist, publishes *A Color Notation.* This book establishes a system of color classification known as the *Color Sphere,* based on the hue, value, and intensity of colors. His basic principles will be taught for a century in schools, illustrated by using his Munsell Color Charts.

Success of Craftsman's Mission furniture (see 1901) now leads to nationally franchised manufacture. This is largely due to designer Harvey Ellis (1852-1904) who joined Gustav Stickley (see 1900) in 1903. But demand soon will exceed production capability, and imitators will abound.

The current Raleigh bicycle in England reaches modern form with pneumatic tires, a diamond-shaped frame, and three-speed gears. Also new this year is the Ladies' Humber bicycle in England, with an open frame to accommodate long dresses.

L.C. Smith & Brothers of Syracuse, New York, introduce its first typewriter. The Royal Typewriter Company is also founded and will produce its first typewriter in 1909.

John Jacques & Son, Ltd., produces a table game using a ball and smooth flat bats. It is called *Ping-Pong* and is distributed by Hamley Brothers, a British game firm that promises "immense excitement and healthy exercise." British engineer James Gibb invented the game in 1890; it was originally named *Gossima.*

01-29

01-26. The Automatic Electric Company incorporates its finger dial feature (see 1901) into its tabletop phone with a new special finger hole for long distance service, now available through AT&T (see 1900).

01-27. Customers love the "hot point" on an electric iron introduced by Earl H. Richardson (see 1903), who developed it last year. The heating elements are arranged to concentrate heat at the point for ironing buttonholes and pleats. It will become the first commercially successful electric laundry iron, and formally be named the *Hotpoint* iron in 1907. In 1912 the company will become the Hotpoint Electric Heating Company.

01-28. National Cash Register Company (NCR) introduces its first electric model, the work of a young inventor hired last year, Charles F. Kettering (1876-1958). Kettering will soon direct an "inventions department" at NCR, and produce a flood of mechanical innovations before his departure in 1912. James Ritty of Dayton, Ohio, invented the original cash register in 1879. He sold his idea to local businessmen in 1884, including NCR founder John H. Patterson (1884-1922). Patterson is credited with inventing the modern corporate structure, creating job descriptions, and establishing a corporate philosophy. The famous office sign "Think," which later will be continued at IBM (see 1911), is already prominently displayed in NCR offices.

01-29. Elizabeth Cochrane (1867-1922) invents and first produces steel barrels (later known as steel drums). Elizabeth is better known as Nellie Bly, the world-famous newspaper reporter who, in 1890, circled the globe in 72 days, beating the record of Jules Verne's fictitious hero, Phileas Fogg (*Around the World in 80 Days,* 1872). Elizabeth headed her millionaire husband's Cast Iron Manufacturing Company upon his recent death, and has established a new steel barrel manufacturing company. Unfortunately, she will go bankrupt and return to newspaper work in 1920.

01-30

01-31

01-32

01-34

1906

Maschinenmöbel (machine furniture) is exhibited at the *Dritte Deutsche Kunst-Gewerbe-Ausstellung* (Third German Applied Arts Exhibition) in Dresden, Germany. Richard Riemerschmid designed the line, and it was produced by Karl Schmidt's *Dresdener Werkstätten für Handwerkskunst* (Dresden Workshops for Handicraft Arts) (see Prologue). A recent article by Friedrich Naumann had inspired Schmidt to develop an innovative fabrication process. The basic box-like furniture pieces are cut in a mechanized serial production process using plain materials that are relatively low in cost.

The *Wiener Werkstätte* (see 1903) holds a *Gedeckte Tisch* (Dining Table) exhibition in Vienna. The press mercilessly pans Josef Hoffmann's silver place setting design (see 1904). Some critics describe the service as "uncomfortable" and "suggestive of anatomical instruments," while others complain that Hoffmann made "geometry, not art" and predict that the style will fail "for practical considerations." Yet the style he innovates will survive in many future forms.

Alfred Carl Fuller (1885-1973) founds the Fuller Brush Company in Hartford, Connecticut. He develops a new machine for making brushes, produces them at night, and sells them door-to-door at 50 cents apiece during the day.

A.W. Faber of Germany introduces *Castell,* a new line of pencils for engineers, technicians, and draftsmen. The pencils are painted green to distinguish them from pencils by their key competitor, L & C Hardtmuth Company of Austria, whose yellow *Koh-I-Noor* pencil became popular after its introduction at the 1893 World's Colombian Exposition in Chicago.

Early electric dishwashers find commercial use. Similar to the one patented in 1886 by Josephine Cochrane, a wealthy Illinois woman, dishwashers remain large and less than effective, with dish breakage as a recurrent problem. Dishwashers will not be accepted in homes until new detergents are developed.

01-30. The Victor Talking Machine Company introduces the *Victrola Model XVI.* Eldridge R. Johnson designed attractive cabinets in several furniture styles that conceal the somewhat unsightly phonograph (still hand-cranked and spring-driven) and its speaker horn inside. Retail price is $200. The elegant and radical concept will transform the phonograph, previously considered a toy, into a popular household item, and the refined design of 1907 will dominate the future industry. See 1901.

01-31. Bissell Carpet Sweeper Company announces that its sweepers are "the only brand advertised nationally to consumers." Invented in 1876 by Anna and Melville Bissell in Grand Rapids, Michigan, more than 3000 per day are now being produced.

01-32. Gustav Stickley of Craftsman Workshops (see 1901) introduces a reclining "Morris" chair. The original reclining chair concept is named after William Morris (1834-1896), who designed a recliner in about 1860

01-34. The Stanley twins, Francis Edgar and Freelan O., set a world's speed record of 127.66 mph. with their Stanley *Steamer* at Ormond Beach, Florida. Their car, developed in 1896, was the first to climb Mount Washington in 1899, and is manufactured by the Locomobile Company of America.

01-35

01-36

01-37

1907

The *Deutsche Werkbund* (German Union of Work) is founded in Munich, Germany, at the urging of Herman Muthesius. The Werkbund aims to marry art and industry, and to develop a new machine aesthetic through an enlightened program of education and improved design for industry. This trade association is the most significant institution created for the progress of industrial design. It consists of 12 industrialists and 12 artists, including Walter Gropius, Peter Behrens, Karl Schmidt, Hermann Muthesius, Henry Van de Velde, Joseph August Lux, Karl Ernst Osthaus, Alfred Grenander, Peter Bruckmann, Friedrich Naumann, and August Endell. Muthesius stated **:** "In *Sachlichkeit* (perfect and pure utility), the problem is to create a new style, a *Machinenstil* (machine style) whose shapes (like bridges and halls) are completely dictated by the purposes they serve."

The *Werkbund* was organized in opposition to inadequacies and the philosophy of the existing *Fachverband für die wirtschaftlichen Interressen des Kunstgewerbes* (Trade Association for the Economic Interests of Applied Art), founded in 1892. The *Werkbund* will be dissolved in 1933 with the collapse of the Weimar Republic but will be re-established in 1947, although never reaching its pre-war prominence.

The *Dresdener Werkstätten*, founded in 1898 by Dresden carpenter Karl Schmidt, and *Münchener Werkstätten für Wohnungs-Einrichtungen* (Munich Workshops for Domestic Furnishings) merge. They become known as the *Deutsche Werkstätten* under Schmidt's directorship.

Lionel Manufacturing Company, founded in 1903 by Joshua Lionel Cowen, markets an electric toy locomotive to go with an electric transformer, and a toy flatcar with motor and battery. The trains run on a three-rail electric-track invented by Cowan and introduced last year.

The Scott Paper Company, maker of toilet tissue, introduces "paper towels," naming them *Sani-Towels.* The product idea resulted from a defective paper mill order of toilet tissue that was excessively heavy and wrinkled. They are sold primarily to commercial establishments.

The Paint Manufacturers Association of the U.S. promotes fashion trends in house paint to match women's fashions.

01-38

01-35. Peter Behrens is appointed artistic advisor to Allgemeine Elektricitäts Gesellschaft (A.E.G.), the successor to DEG (Deutsche Edison Gesellschaft), a German company founded by Emil Rathenau in 1883 to produce Edison's light bulb. Behrens will become intensely involved with architectural, product, and graphic design programs (such as this poster) at AEG until the war. Behrens said, "It is agreed; we refuse to duplicate handmade works, historical style forms, and other materials for production."

01-36. The Parsons Band Cutter and Self-Feeder Company of Newton, Iowa (later known as Maytag) debuts this *Pastime* washing machine as a side business to its farm equipment. Frederick L. Maytag has headed the company since 1893. The clothes washer was designed by pattern maker George Seed with a corrugated wooden tub and a hand-operated dolly inside, but will not acquire electric power until 1911.

01-37. The new The Automatic Electric Company's wall model telephone is very compact, and advertises the Strowger patented telephone exchange on the dial.

01-38. Jacob & Josef Kohn, a competitor of Thonet, introduces the *Fledermaus* armchair #728, with decorative wood spheres and horseshoe-shaped base designed for the interiors of the *Kabarett Flederma,* designed by architect Josef Hoffmann. It will be widely copied and re-introduced in 1980 by Thonet.

01-39

01-40

01-41

1908

Sears, Roebuck & Company (founded 1893) begins to sell ready-to-build house kits and plans, ordered by mail and delivered by railroad. They will continue to sell until 1940.

Viennese architect Adolf Loos writes his essay, *Ornament und Verbrechen* (Ornament and Crime), urging the elimination of ornament from buildings and products. He argues that what was necessary in the past is no longer appropriate in a modern industrialized world, and that the greater the decoration on an object the greater the exploitation of the craftsman employed to create it.

The *Deutsche Werkbund* (see 1907) finances a new museum in Hagen, Germany, called the *Deutsches Museum für Kunst in Handel und Gewerbe* (German Museum for Art in Trade and Commerce). It will be used to disseminate information on German design through its own exhibitions (see 1914), and as a base for traveling exhibitions throughout the world, including the U.S. (see 1912).

The American Home Economics Association is founded to raise the status of housewives, and improve the conditions of housework through research, education, and the application of the newest science and technology to the home. Its first president is Ellen Swallow Richards, and publication of its *Journal of Home Economics* begins. Richards states, "Tomorrow, if not today, the woman who is to be really mistress of her house must be an engineer, so far as to be able to understand the use of machines."

A 15-year old French boy, Raymond Loewy, designs and builds a model airplane powered with rubber bands. He names it the *Ayrel,* patents it, forms a company, and rents space for production. He will become a pioneer industrial designer in the U.S. See 1933.

Ford introduces the Model T, available at $845 in black only (similar cars cost $2000 to $3000). It is affectionately called a "Flivver" (a small, old, or cheap automobile) or a "Tin Lizzie," and is said to be "stronger than a horse, and easier to maintain." In 1909, Henry Ford will declare that he will build only this model in the future. Before its end (see 1926), 16 million will be produced.

01-39. Peter Behrens designs two new trademarks for AEG (lower left and upper right), replacing older marks designed by Franz Schwechten in 1896 (upper left) and Ludwig Sütterlin in 1894 (middle) By 1914, Behrens will design the trademark still recognizable in 1995 (lower right).

01-40. The Hoover Electric Suction Sweeper Company introduces its first vacuum cleaner, *Model 0,* the first commercially successful portable electric vacuum cleaner produced in quantity. The device was patented this year by James Murray Spangler, an asthmatic "cousin-in-law" of William H. Hoover, an Ohio maker of leather harnesses and equipment since 1875. In 1922, the company will be renamed The Hoover Company.

01-41. Peter Behrens designs this table-fan that is marketed by AEG (see 1907). It will be produced in several fixed and swivel models over the next few years.

01-42. General Electric introduces its first widely distributed electric toaster, adding to its already full line of consumer heating and cooking devices. The two-slice D-12, with a porcelain base, will remain in production until 1913.

01-42

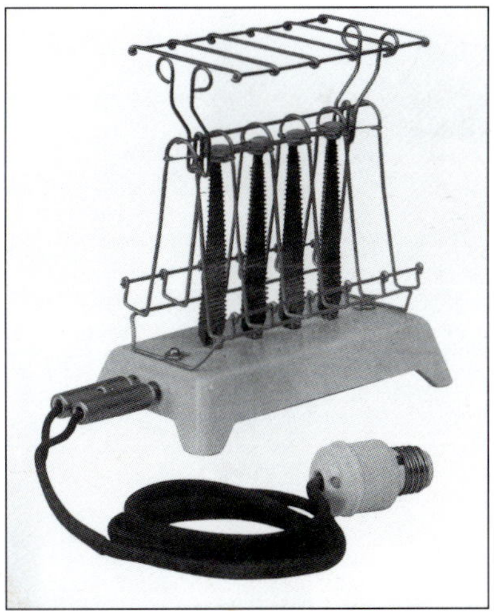

01-43

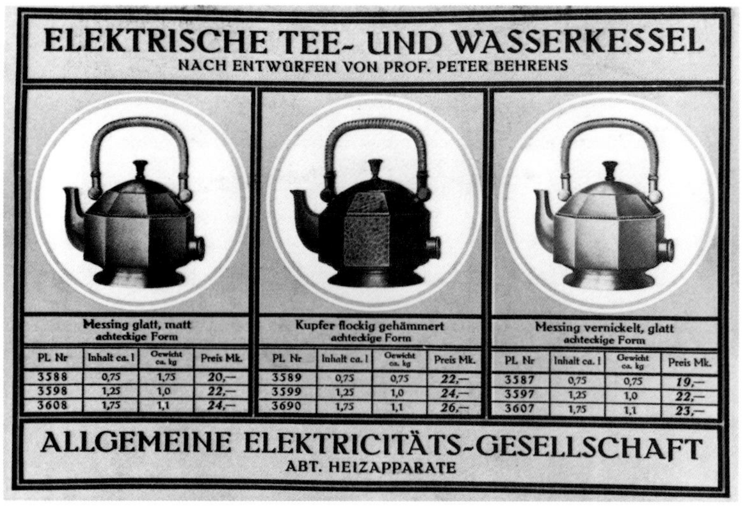

01-44

01-45

01-46

1909

Poet Filippo Tommaso Marinetti (1876-1944) launches *Futurism,* an Italian fine arts movement that is the first to be directed at a mass audience, with a manifesto. It aims to celebrate the machine and global communications now changing the world. Marinetti hailed war as "the world's only hygiene...the destructive gesture of freedom-bringers." The movement will be strengthened when it is joined by, and becomes the expression of, architect Antonio Sant'Elia. Many Futurists will embrace Fascism, designing nationalistic symbols and propaganda when Mussolini reaches power in Italy in 1922.

The *Gödölló Workshops* in Hungary (see 1902) hold a group exhibition at the National Salon in Budapest. Its work features vernacular furniture and weavings derived from the Arts and Crafts movement, and focused on moral or "national" themes. Highly acclaimed at the 1904 Louisiana Purchase Exposition in St. Louis, Missouri (see 1904), and state-supported since 1907, the experiment will collapse in 1921.

Dr. Leo Hendrik Baekeland (1863-1944), U.S. chemist and inventor of Kodak's *Velox* photographic paper, announces a phenolic plastic called *Bakelite.* He invented, developed, and patented the plastic starting in 1907, hoping to find a substitute for shellac, which was used to insulate electrical wires. *Bakelite* is the first synthetic plastic, a thermosetting resin of phenol and formaldehyde, reinforced with finely ground sawdust or flour, and used initially in electrical components. He founds the General Bakelite Corporation to produce the material that will become popular for consumer product housings from the early 1930s until after World War II.

German housewife Melitta Bentz exhibits her new filter coffeepot at the Leipzig, Germany trade fair. She has designed a coffee maker that uses discs of porous paper to filter-drip the coffee. By 1912, she will be manufacturing her own line of paper filters, but the patented filter drip method will not reach the U.S. until 1963.

The H.G. McFaddin Company of New York introduces *Emeralite* electric desk lamps. The lamp's green glass shade lined with white opal glass will be imitated extensively, and will be produced into the 1940s.

01-43. Peter Behrens completes this Turbine Factory Hall in Berlin for AEG in Germany. It is a strong architectural expression of the concentrated power, order, precision, and regularity of production methods. The corner pillars and roof construction suggest classicism, but the steel-supported building has glass surfaces so that the structure remains visible.

01-44. Peter Behrens designs electric water kettles in copper and brass; they are produced by AEG in Germany. The design allows for eighty variations of three basic forms: a cylinder, half-oval, and octagonal, each in three sizes, although only thirty were produced. AEG production engineer Michael von Dolivo-Dobrowolsky conceived this use of mass-produced standardized components to create a wide variety of products.

01-45. Since 1906 Roycroft (see Prologue) has produced this small-scale chair No. 035, originally designed for Marshall P. Wilder, a little person. The chair is produced in quantity as a dressing table chair or an intermediate size between a baby's high chair and an ordinary dining room chair.

01-46. Craftsman Workshops (see 1901) introduce this oak settle designed by Gustav Stickley.

Edison

Chapter 2

1910-1919 The "Modern" Movement

Driven by the technologies and mass marketing of automobiles, airplanes, communication, and electricity, the U.S. creates a consumer-based standard of living which transforms the Victorian arts and crafts mentality into one of industrial mass-production.

The national standardization of electric current in 1910 allows proliferation of electrical machines and appliances. Few of these products, if any, are concerned with appearance, but, for mechanized farms and homes that previously operated by only animal or human manual labor, such tools provide profitability and new leisure time.

The landscape bristles with tall urban buildings. Stark but powerful, smoking factories produce ever-increasing quantities of products at lower cost. New manufacturing processes created for Henry Ford's Model T in 1913, and efficiency studies by Frederick Taylor and his followers from 1911 onward, increase employment opportunities, purchasing power, and domestic markets, and initiate a mass migration from rural to urban living.

Trends in commerce influence everyday life as well. In 1912, Christine Frederick popularizes application of industrial principles to the domestic scene as "Home Economics". The American home itself is being transformed. Frank Lloyd Wright's publications in Germany impress a new generation of European architects, and merge with minimalist theories of Loos, Le Corbusier, geometric cubism and abstract art.

German architects embrace the functional style known as "factory aesthetics" at the same time that Albert Kahn is designing the new Ford manufacturing complex in the U.S. A *Werkbund* exhibition in the U.S. acquaints Americans with new aesthetics for home products and furnishings. The New York Armory Show of 1913, with "modern" art by Duchamps, Picasso, and others, shocks the American public.

By 1914 there are a half million cars in the U.S., and traditional farm life patterns are affected by commuting and shopping. Mobility creates suburbs that encourage early shopping center developments and self-service retailing. *Victrolas*, *Pyrex* , telephones, and refrigerators are examples of the wondrous new consumer products generated by corporate research.

While artistic traditions of the dying Victorian age dominate the 1915 Panama-Pacific Exposition in the U.S., European art continues its headlong revolt against tradition. Kasimir Malevich pioneers abstract art in 1915, and the Dadaists in 1916 discard even the traditional materials of art. In 1917, *De Stijl* in the Netherlands advocates the abandonment of natural forms.

The First World War devastates European economy and culture, but also develops new technologies and materials, strengthens U.S. manufacturing capabilities, and seeds U.S. design culture with young European design émigres who contribute significantly to the modern movement here.

In post-war Russia and Germany, industrial models of "Fordism" and "Taylorism" from the U.S. influence economic recovery programs. Constructivist artists in Russia look to factories for inspiration. In Germany, the Bauhaus is established in 1919 to train designers for high-quality export goods.

While artists, architects, and designers still struggle to label and express it visually, the post-Victorian, as yet unnamed phenomenon of the "Modern" industrial movement, represented by images of smooth, machine-like, geometric surfaces, is now the face of the 20th century. The U.S. joins the rest of the world as a generator of progressive design thinking.

02-01

02-02

1910

Françoise Spoturno (better known as Coty) of France revolutionizes perfume packaging by asking jeweler and glassmaker René Lalique (1860-1945) to create quality decorative containers. Lalique, famous for nineteenth century jewelry design, abandoned this work last year upon the death of his wife, Augustine-Alice Ledru, and turned to cast glass as his new medium. His designs will enliven perfume bottles, statuettes, and lighting fixtures.

A complete porfolio of the work of Frank Lloyd Wright, *Ausgefürte Bauten und Entwürfe von Frank Lloyd Wright*, is published in Berlin by Ernst Wasmuth. The book describes more than 130 buildings. A second, less-lavish edition will be published in 1911 with an introduction by C.R. Ashbee (1863-1942), whom Wright has known since 1901. Through these publications, Wright's work will become well known to the younger generation of architects in Europe.

In Chicago, the Chicago Flexible Shaft Company, producer of sheep-shearing machines and hand clippers, enters the home appliance field with the *Princess* electric iron. In 1921 the company will first use the name *Sunbeam* to brand its products, and this will become the name of the company in 1946.

Fred Osius, L.H. Hamilton, and Chester A. Beach of the Arnold Electric Company in Racine, Wisconsin, produce an electric eggbeater. Since 1904 Osius and George Schmidt had been making and selling electric stirrers for soda fountains to beat ice cream and milk into a "shake." In 1905 Beach developed the first high-speed (7,500-rpm) electric motor, and installed it in a vibrator sold to the barber supply trade. Next year Arnold will change the company's name to the Hamilton Beach Manufacturing Company, and will shortly introduce a new "Universal" motor that will empower many household appliances such as sewing machines. Drink mixers will follow and, by 1918, the motor will be adapted for food mixers.

James Kirby of Cleveland, Ohio invents a vacuum cleaner with cleaning attachments, Model B. Produced by three brothers, Edward L., Clarence G., and Walter Frantz, it is marketed under the name of Frantz-Premier. The brothers will leave to set up their own vacuum cleaner company, the Apex Electric Manufacturing Company, in 1912, and the Kirby cleaner will debut after World War I, when James Kirby associates with the Scott & Fetzer Company.

02-01. The Hughes Electric Heating Company in Chicago, just founded by George A. Hughes (1871-1944), introduces the first "electric cook stove," with three burners and an oven, at the National Electric Light Association convention in St. Louis, Missouri. In development since 1904, his cook stove is finally produced in 1909**.** Earlier cooking "tables" were merely a series of electrical plugs for individual accessories. See 1918.

02-02. The Victor Talking Machine Company introduces this Model XI *Victrola.* A design patent for what will become the typeform of phonograph cabinets is issued to John C. English. More than a million will be made through 1926.

02-04

02-06

1911

Philadelphia engineer Frederick Winslow Taylor (1856-1915), publishes *Principles of Scientific Management.* He seeks to design a totally efficient system of production, and founds the Taylor Society, which will establish “industrial engineering” schools throughout the country. Taylor had studied work processes as chief engineer at the Midvale Steel Works (1880-1900). His system, known as *Taylorism*, will become a description of the American system of mass production, and will motivate “efficiency experts” with stopwatches to study businesses of all types. One of his students, Henry L. Gantt will develop “flow charts” that allow management to track the progress of product development. Other apostles will be Frank Gilbreth (see 1914) and Christine Frederick (see 1912)

General Electric markets the *Audiffren,* an automatic electric sulfur dioxide compressor refrigerator, with a wooden enclosure. Invented in France by Abbé Marcel Audiffren, a French monk, it is produced under contract to the American Audiffren Company. It sells for $1000, twice the price of an automobile, and is the first refrigerator for household use in the U.S. It will be manufactured until 1928.

Thomas J. Watson Sr., a former executive of the National Cash Register Company, founds the Computing-Tabulating-Recording Company in a merger with Herman Hollerith’s Tabulating Machine Company. Hollerith invented his punch-card tabulating machine for the 1890 U.S. census. The new company will make time clocks and scales, and will be known as International Business Machines (IBM) after 1924.

In lectures, “Monumental Art and Industrial Building,” Walter Gropius shows concrete silos in three cities, Baltimore, Ohio, Buffalo, New York, and Minneapolis, Minnesota, to illustrate forms expressing the monumentality and power of American technology and civilization. Gropius states, “The artist possesses the ability to breathe soul into the lifeless product of the machine.”

The Upton Machine Company is founded in St. Joseph, Michigan to produce electric motor-driven wringer washers, branded as “Whirlpool” washers. By 1916, Sears, Roebuck and Company is selling Upton-manufactured washers, under the name of "Allen," faster than Upton could make them. Upton will become the Nineteen Hundred Corporation in 1929, and the Whirlpool Corporation in 1950.

The Kohler Company introduces the first bathtub with a decorative skirt that reaches the floor. The new style will gradually replace the present skirt-less tubs that look like their original design inspiration—a metal water trough for livestock.

02-05

02-04. This Faguswerk (Fagus Factory), a shoe-last manufacturing facility in Alfeld-an-der-Leine, Germany, is designed by Walter Gropius and Adolph Meyer using glass and steel for exterior walls and the first of what will be defined as “curtain walls.” In the 1950s, it will be designated for landmark preservation in recognition of its architectural significance.

02-05. Thomas Edison’s National Phonograph Company introduces this Model F Standard Phonograph with a “cygnet” horn. Edison’s is one of the few companies still using the now outdated cylinder system Edison invented in 1877.

02-06. Washington A. Roebling II, grandson of John A. Roebling, builder of the Brooklyn Bridge in 1883, produces the Type 35 Mercer *Raceabout* automobile, designed by Finley Robertson Porter. Washington wants to lead the staid Roebling firm into the new age of the horse-less carriage, but the firm will fade when he is lost on the *Titanic* in 1912, and will be sold in 1919.

02-08

02-09

1912

Deutche Werkbund (see 1907) sends *Modern German Applied Arts,* a 1,300-piece exhibition of its work to the Newark, New Jersey, Museum of Art and other U.S. institutions. John Cotton Dana, founder of the Newark Museum in 1909, and its director, sees the Museum as a department store for modern ideas of good taste. It will later hold an annual event called *Inexpensive Articles of Good Design* with items under 50 cents from local stores.

In Paris, Paul Irigbe completes the first interior design in the style that will become known as "Art Deco" for couturier Jaques Doucet, to house his collection of cubist and surrealist paintings.

Christine Frederick publishes *The New Housekeeping,* a series of four articles in *Ladies Home Journal*, which begins to redefine the home and will be published in book form next year. She was inspired by her husband's work with Frederick Taylor (see 1911) and an earlier book, *The American Woman's Home* (1869), by Catherine Beecher and her sister, Harriet Beecher Stowe. See 1915.

An exhibition in Berlin, *Women at Home and at Work*, features furniture designs by Lotte Klopsch. Arthur Schubert, a Dresden furniture showroom owner inspired by the *Deutsche Werkstätten* (see 1907) hired her to design furniture. She is the first designer in a new association of furniture sellers, later to be named *Deutsche WK* (*Wohnungs Kunst*) *Möbel* or "German Art Furnishings") which will commission its own designers to create a unified style.

In the U.S., Anna Wagner Keichline (1889-1943), a recent architectural graduate from Cornell University, patents a new multiple use kitchen sink and drain board arrangement featuring an efficient use of space and user convenience. Keichline, already an accomplished cabinetmaker at age 14, will devote her talents over the next 20 years to "industrial art" designs aimed at improving the comfort, efficiency, appearance, and convenience of the kitchen and other household furnishings and interiors.

Gorham Manufacturing Company, Providence, Rhode Island, makes a silver table service, designed by architect George Washington Maher (1864-1926) as part of the furnishings in his Rockledge House in Homer, Minnesota. The pattern uses Gorham's hand-wrought *Martelé* finish, a shimmering small hammer mark effect designed by William Christmas Codman in 1897. In 1989, Dansk (see 1954) will purchase Gorham.

02-10

02-08. Roycroft Workshops "Morris" chair No. 043.

02-09 . Glenn Hammond Curtiss (1878-1930), of the U.S., builds and successfully flies this, the first practical seaplane or flying boat, the Curtiss E Type. Curtiss began his experiments with his land plane mounted on floats in 1910, and will soon receive the first naval contract for two of his seaplanes.

02-10. The American Electrical Heater Company of Detroit (founded 1894 by Ben Scranton) introduces the Model 6 1/2 B *American Beauty* household iron, which will set the product typeform. It takes its name from its weight, 6.5 lbs. Four million will be produced before production ends in 1930.

02-11

02-13

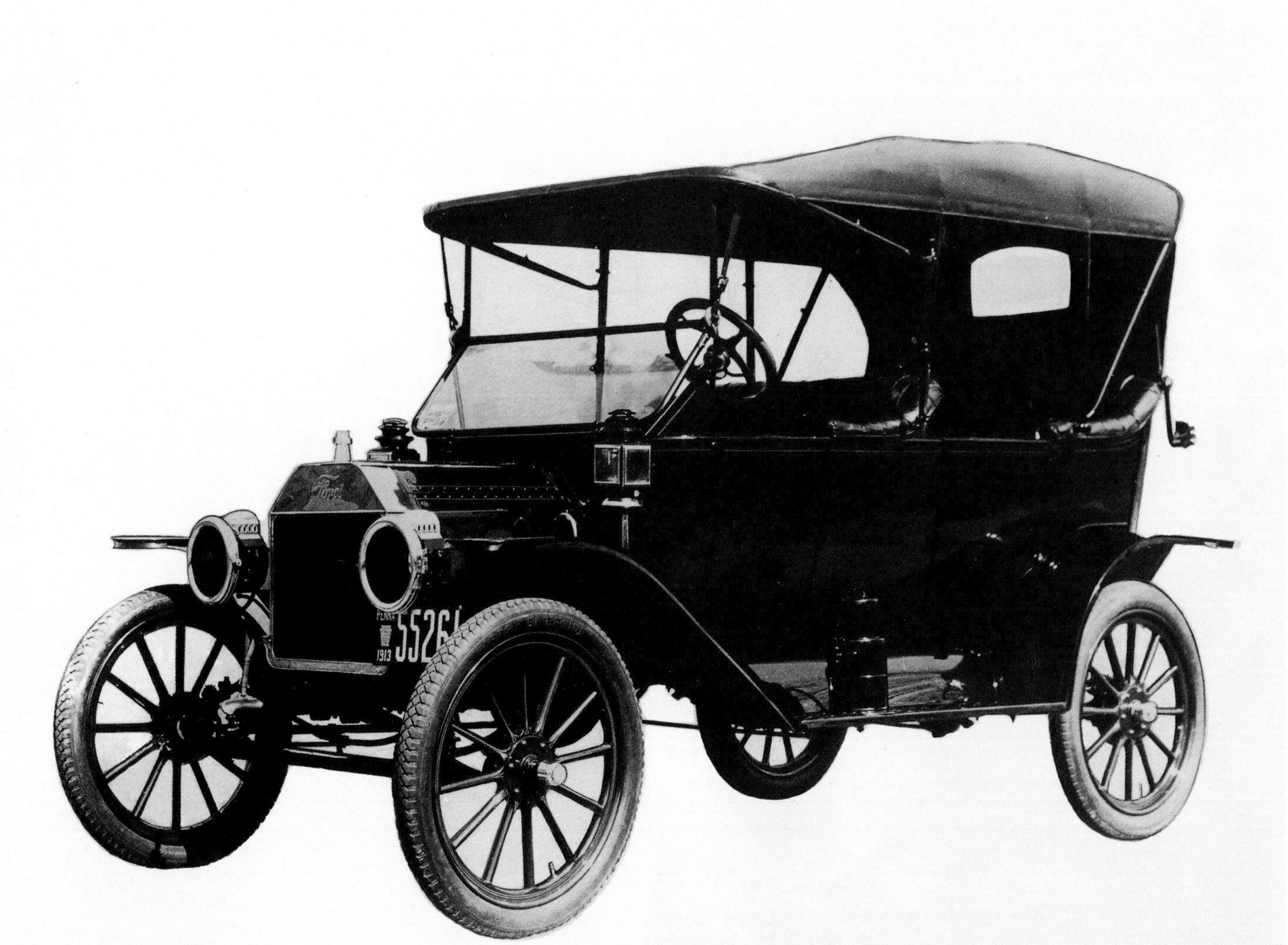

1913

Edward B. Moore, U.S. Commissioner of Patents, urges the modification of existing regulations to protect the property of "industrial design, the distinguishing form of products that have marketable value." This is the first known public use of the term, which generally was used as a synonym for "art in industry". It will not be commonly used much until the mid-1930s, when the new profession promotes the term.

The *Kewpie Doll,* designed and patented by New York author/illustrator Rose Cecil O'Neill, is made of celluloid from molds prepared by Pratt Institute student Joseph L. Kallus. It will earn $1.5 million for O'Neill.

The first *Erector Set*, by inventor/designer/producer Alfred Carlton Gilbert (1884-1962), is introduced. Made of steel and designed in 1911, it was inspired by steel girders used in the electrification of the New Haven Railroad. Gilbert, a medical doctor, magician, and former Olympic pole-vaulter, produces the toy at his own Mysto Manufacturing Company which will, in 1916, be renamed the A. C. Gilbert Company.

The Formica Corporation is established, and begins producing a paper-based laminated material used for electrical insulation as a substitute "for mica," hence the name, *Formica.* Dr. Leo Baekeland (see 1909) obtained the initial laminating patent in 1911, and was joined in its development by Westinghouse engineers Daniel J. O'Coner, Jr. and Herbert A. Faber. By the 1920s black and brown *Formica* laminates will be used in the furniture industry, and by 1927 a method for printing decorative wood grain patterns on the surface will be patented. See 1950.

02-12

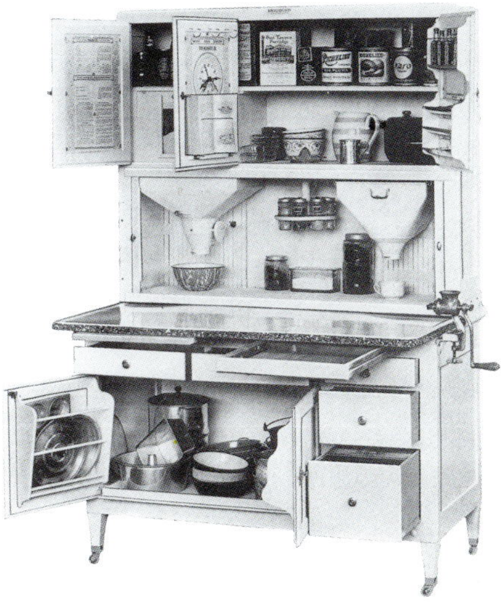

02-11. The Walker Brothers electrify their dishwashing appliance, thus creating the first "automatic" dishwasher. They created the original 1909 mechanical model in the rear of their hardware store in Syracuse, New York The dishwasher was invented in 1888 by Josephine Garis Cochrane (1839-1913), with hand-pumped soapy water over dishes in wire racks. Her company, The Garis-Cochran Dish-washing Machine Co., will be absorbed by Hobart (see 1919) in 1926, later to be owned by the Whirlpool Corporation, which will regard Cochrane as its founder.

02-12. These "Hoosier" kitchen cabinets, with designated places for storage of flour, sugar, and spices and working surfaces for kitchen tasks, are at the height of popularity. Made by the Hoosier Manufacturing Company since its 1899 founding in Albany, the freestanding units will remain important until built-in cabinets change the organization of the kitchen in the 1920s.

02-13. Henry Ford perfects the mass-production process with this, his redesigned car, the Model T. Inspired by efficient Chicago meatpacking processes, Ford develops a sophisticated assembly-line method that reduces production time from 12 1/2 to 2 1/2 hours, and the car price drops from $900 to $440. This year alone, 168,000 Fords are produced. His unprecedented system will become known throughout the world as *Fordism*, and by 1915 will reduce skilled labor in auto factories by nearly 40 percent. Fifteen million Model T's will be produced and almost as many jokes about the "Flivver" (a cheap, old, or small automobile) appear, but in 1960 designer Jay Doblin will deem it "the most beautiful car ever built, a classic illustration of what an automobile should be." The actual designer, engineer Harold Willis, would be proud.

02-14

02-15

1914

Frank Gilbreth publishes *A Primer of Scientific Management.* An apostle of Frederick Taylor (see 1911), he uses time-and-motion studies to break down labor into 18 basic movements he calls "Therbligs" (his name backwards). Unlike Taylor, Gilbreth is concerned that the worker remains a person, not a cog in a machine. His wife Lillian Gilbreth (1878-1972) will write several books on domestic management and develop the concept of the kitchen "work triangle," the efficient arrangement of sink, stove, and refrigerator. Their life with 12 children will be humorously chronicled in a 1948 best seller and 1950 film, *Cheaper by the Dozen.*

The *Deutsche Werkbund* (see 1907) holds an exhibition in Cologne, Germany. It includes an all-glass pavilion by Bruno Taut, and two designs by Walter Gropius and Adolf Meyer, an office building and a machine hall, which were influenced by Frank Lloyd Wright's bank and boat club designs in his Wasmuth publications (see 1910). The exhibition establishes the "factory aesthetic" which will be incorporated into the modern movement. The buildings will be demolished after the exhibit. At this exhibition a debate on official national design policy rages between Henry Van de Velde, who favors design individuality, and Hermann Muthesius, who favors bureaucratic standardization of designs. Walter Gropius joins with Van der Velde in an unsuccessful attempt to seize control of the *Werkbund*, but the war intervenes and silences the debate.

The "Hookless Fastener" is first marketed. It was perfected by engineer Otto Frederick Gideon Sundback (b.1880) for the Hookless Fastener Company in Meadville, Pennsylvania, during the last eight years, and is based on an 1893 patent by his predecessor, Whitcomb L. Judson, as a "Clasp Locker or Unlocker for Shoes." It will later be named a *Zipper* (see 1923).

Cellulose acetate, made from vinegar, acetic acid, and cotton, is invented in Germany, and used as a skin coating for fighter aircraft.

The Kwaishinsha Company in Japan produces its first car, the DAT (Japanese for "hare"). Masujiro Hashiumoto founded the company in 1911. In 1930, it will produce the Datson (son of DAT), but since "son" in Japanese means "ruin" it will be changed to Datsun. After 1935 the company will be known as Nissan.

Tinker Toys are introduced. Their basic spool and dowel building components will remain basically unchanged through the century.

02-16

02-14. This, the first Dodge car, is introduced by John and Horace Dodge, using all-steel body construction developed in 1912 by Edward G. Budd Manufacturing Company instead of traditional steel-over-wood. This all-steel construction is the first in the industry. The Dodge's immediate success as a quality product will inspire widespread imitation by all manufacturers and allow engineers to standardize body designs.

02-15. The Stutz *Bearcat*, one of the most famous early American sports cars, designed by Harry C. Stutz (1876-1930) debuts.

02-16. A.H. Goss, an executive, and E. J. Copeland, a purchasing agent, both of General Motors, contract independent inventor Nathaniel B. Wales to develop this domestic refrigeration machine. Although he had originally worked on an absorption process, Wales settles on a compression machine using sulfur dioxide as a refrigerant after several test models. See 1918.

02-17

02-18

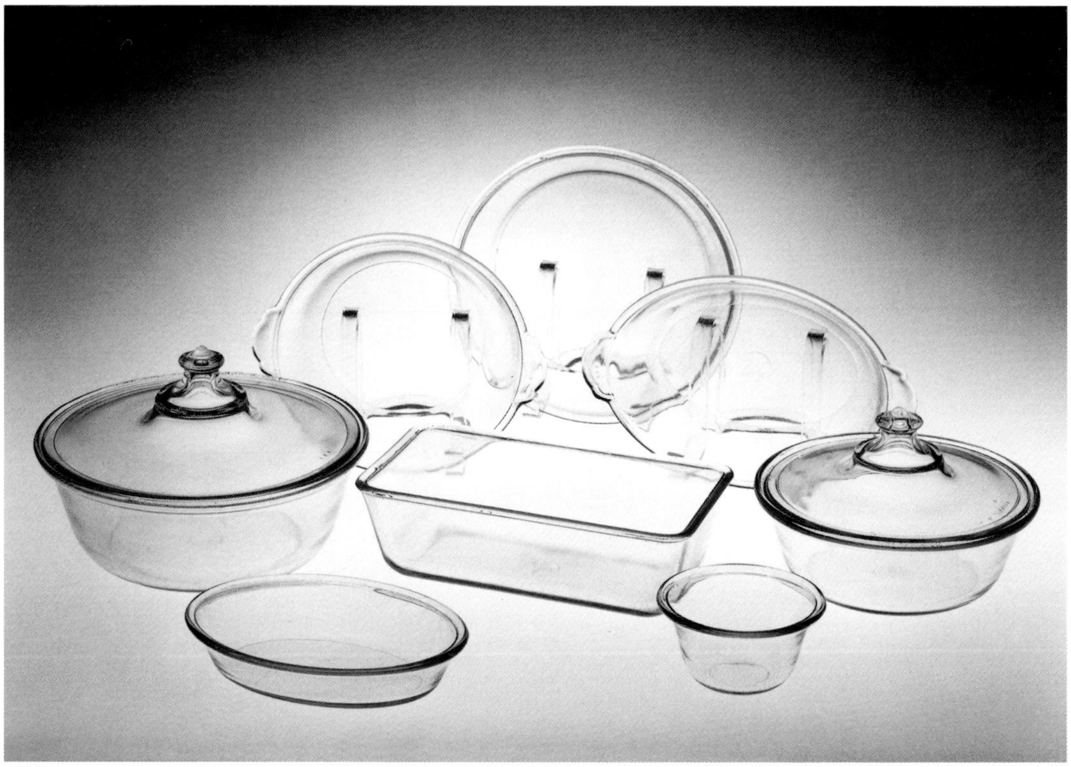

02-19

1915

The Design and Industries Association (DIA) is established by Cecil Brewer and younger members of the Arts and Crafts Exhibition Society in London. The DIA, modeled after the organization of the *Deutsche Werkbund* (see 1907), does not endorse modernism, but is an important forerunner of the later Council of Industrial Design (see 1944).

The Color Association of the U.S. begins the forecasting of colors based on the growing popularity of color in consumer goods.

Stove manufacturers begin to shift from cast iron to enameled steel surfaces.

Christine Frederick (see 1912) publishes *Household Engineering*, a book that recommends principles of efficient home design: reduced floor space, built-in furniture, and a closely related sink, icebox, and work area. She illustrates her ideas with a new home built in Tracy, Illinois.

Mary Pattison publishes *The Principles of Domestic Engineering*. She has spent several years applying Frederick Taylor's method (see 1911) to home activities, and has established a "Housekeeping Experiment Station" in Colonia, New Jersey.

The U.S. Postal Service introduces a standardized rural mailbox, selecting the design of one of its own draftsman, Roy Joroleman, after a 14-year study by a commission named to resolve the problem of design proliferation. The design will become known as the "breadloaf."

The Sopwith *Camel* is a British scout and fighter plane designed by Thomas Sopwith, a former racing pilot who won the Schneider Trophy with his *Tabloid* racing plane in 1913. After the First World War, Sopwith's company will become H.G. Hawker Engineering Company, Ltd., and later, Hawker Aircraft Ltd.

02-17. Alex Samuelson, the plant supervisor for the Root Glass Works of Terre Haute, Indiana, develops a new Coca-Cola bottle shape. It is called the "hobble skirt" design after current ladies fashions and will be introduced widely in 1916. This now-familiar contour shape will be granted registration as a trademark in 1977, and a modern contour bottle will be introduced in 1993 in a 20-ounce plastic version.

02-18. Corning Glass Works introduces the first glass ovenware, made of a new, clear, heat-resistant material, *Pyrex®*. Corning chemist Dr. Jesse T. Littleton discovered it in 1913 when he provided his wife with a makeshift casserole made from a cut-down *Nonex* battery jar. This experiment survived the oven as well as traditional ceramic casserole dishes. Corning revised the formula to remove the lead that is dangerous in food products. The first item produced was a circular nine-inch "pie" plate. The name *Pyrex ®* was derived from the Greek word "pyra" or hearth.

02-19. The Panama-Pacific International Exposition is held in San Francisco, honoring the opening of the Panama Canal. Characterized by spectacular night illumination of 48 searchlights, the Exposition celebrates the arts, include such buildings as the *Palace of Fine Arts* by architect Bernard Maybeck (illustrated), *The Court of the Universe* by architects McKim, Meade and White, and paintings such as *The Pioneer* by Solon Borglum and *The End of the Trail* by James Earl Fraser. A concurrent Panama-California Exposition in San Diego, promotes resources and opportunities for development and commerce of the western states.

02-21

1916

The *Normenausschuss der Deutschen Industrie* (German Standards Commission) is formed. It will work to produce national standards for products of all types under the title of *Deutsche Industrienormen* (DIN-or German Industry Standards).

The Arts and Crafts Society of Sweden begins to encourage the owners of glass, ceramic, and other factories to hire artists and designers to elevate the national level of product design. This group will encourage many artists to create beautiful things for everyday use, and will lead to what will be called "Swedish Modern." See 1919.

Frank Pick (1878-1941) implements an impressive corporate identity system for the underground London Transport. The graphics program features a bar and circle trademark, and a display typeface to be used throughout the system, designed last year by calligrapher Edward Johnston (1872-1944). Pick was a founder of the Design and Industries Association (see 1915), and will become managing director of London Transport from 1928 until 1940.

Eversharp (U.S.) introduces its first mechanical pencil, produced by the Wahl Company in Chicago, Illinois. Some say that last year Japanese engineer Tokuji Hayakawa marketed a mechanical pencil under the brand name Sharp, and used the proceeds to establish the Hayakawa Electric Company in Japan, which will market a variety of products under the Sharp name.

Several new refrigerator production companies are founded this year. In May, the Goss & Copeland Electro-Automatic Refrigerator Company is formed. The new company is almost immediately renamed Kelvinator, and will shortly introduce its first product for household use. Another company developing such a device, Frigidaire, is also founded this year, and will be purchased by General Motors in 1919.

Kark Rapp and Gustav Otto found the Bayerische Motoren Werks (BMW) to manufacture aircraft engines. In 1923 they will manufacture motorcycles and, in 1929, automobiles. The famous blue and white trademark will not be used until then, but its design will symbolize a spinning aircraft propeller, and use colors currently used by Bavarian Luftwaffe planes.

02-20

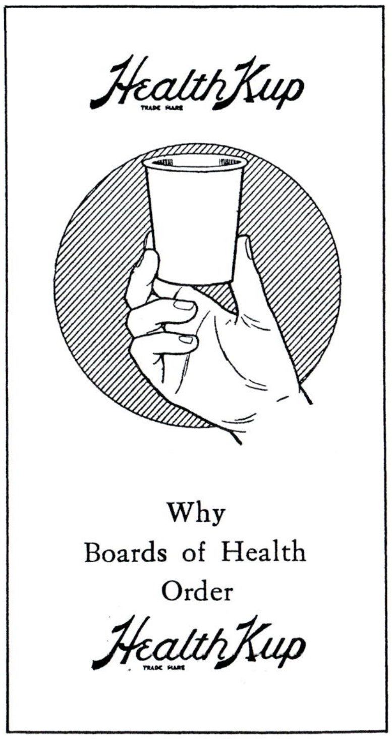

02-20. Contracts for *Health Kups*, disposable paper cups, are sold to over 100 railroads throughout the U.S. They are manufactured by the Public Cup Vendor Company, founded by Hugh Moore and Lawrence Luellen in 1909. Luellen developed the disposable cups because public health officials warned that disease was transmitted from cups shared by many people in public places. Charles M. Gage Jr. patented the final design in 1912. In 1919, after the influenza epidemic increases demand enormously, Moore will change the company name to the Dixie Cup Company to distinguish it from competition. The product will be known as a "Dixie" cup, named after the Dixie Doll Company of New York, a manufacturing neighbor and friend of Moore. In 1923, the company will manufacture most retail ice cream containers, as well.

02-21. The military uses this rugged chain-driven, solid-tire truck built by the Mack brothers, wagon builders who built their first truck in 1906 in Allentown, Pennsylvania. Nicknamed the "Bulldog" by soldiers, the bulldog symbol will become a Mack trademark.

02-23

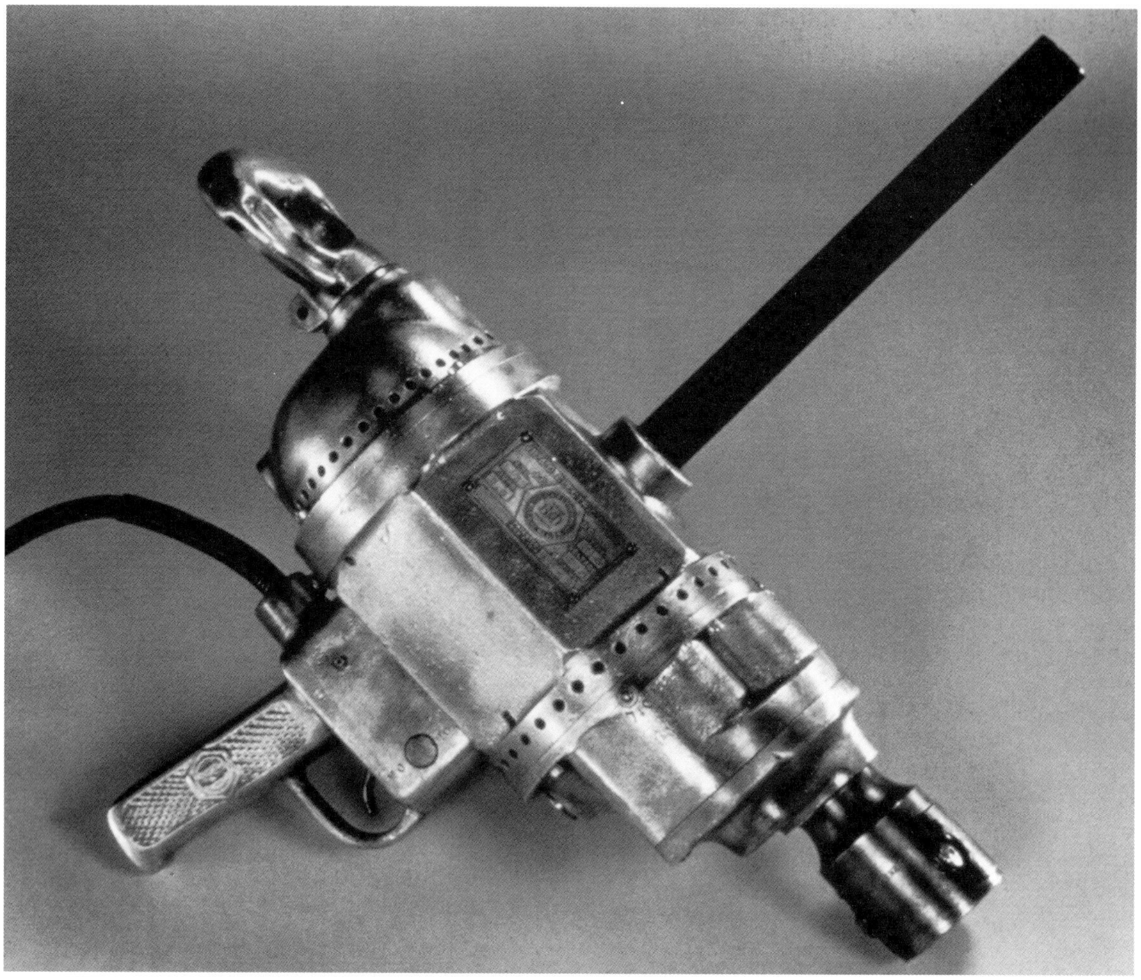

02-24

1917

Scottish zoologist D'Arcy Wentworth Thompson (1860-1948) writes a treatise, "On Growth and Form." He uses the term "stream-lining" to describe organic structures that offer the least resistance when they are in motion, and how a fluid medium tends to impress its "stream lines" on deformable bodies such as snowdrifts and sand dunes. He concludes that the same principle must have come into play in the evolution of the bodies of fish and birds. The term "streamlining" was originally coined in 1893 in connection with hydrodynamics, but now becomes popular to describe the styles of cars and trains.

De Stijl, a journal by Dutch sculptor and architect Theo van Doesburg, is published to promote the modern art movement of the same name he founded in Leyden, the Netherlands, in 1914. He advocates the use of primary colors, asymmetrical balance, and the abandonment of natural forms in architecture and art. Leading figures associated with the movement are the painter Piet Mondrian, the architect J.J.P. Oud, and architect Gerrit Rietveld. They will issue a manifesto next year, and their publication *De Stijl* will continue until 1928, but the movement will fade with the death of van Doesburg in 1931.

Bolsheviks seize power in Russia, and believe that modern American Fordism (see 1913) and Taylorism (see 1911) can lead the way to a Socialist future. Based on this premise, *Constructivism*, a new artistic movement begins; it urges artists to "go into the factory, where the real body of life is made," and link art to mass production and a new social order. The movement will exert strong influence on the modern movement in the West, and will last until about 1922, the year the USSR is formed. See 1920.

The Converse Rubber Company debuts its "All-Star" sneakers, which in 1923 will be renamed the "Chuck Taylor All-Star," after basketball pro Taylor endorses the shoes. They will be emblazoned with a distinctive round ankle patch with a star. U.S. Rubber Company calls its new product "U.S. Keds." Such shoes have been on the market since the vulcanization process of bonding rubber to cloth was invented in the 1860s, but were simply called "sneaks" or "sneakers" in the U.S., or in England, "plimsolls."

02-23. The first electric hand-held drill with a trigger switch, designed and introduced by S. Duncan Black and Alonzo G. Decker last year, is granted a patent. Black & Decker Manufacturing Company was founded in Baltimore, Maryland, in 1910, and makes a variety of specialized machinery and equipment.

02-24. The Fokker Dr 1 triplane enters war service for Germany. One of the first to use it is Manfred von Richthofen, the "Red Baron" ace who paints his plane all red to identify himself to the enemy. He commands the famous "Flying Circus" squadron, so called because its aircraft were painted in distinctive, bright colors.

02-25. A typical contemporary "trolley car." There are over 1000 electric street railways in operation in the US, with nearly 45,000 miles of track. They originated in 1888 with low cost electric power and F.J. Sprague built the first practical system in Richmond, Virginia.

02-25

02-26

02-28

1918

Konosuke Matsushita founds the Matsushita Electric Industrial Company in Japan, and states, "I decided that the task of the industrialist is to make his products available at the lowest possible cost to bring better living to the people of the world." His first product is a bicycle lamp. The company will later become known for Panasonic, Technics, and Quasar brand names.

Ernst Keller (1891-1968) joins the faculty of the Zurich School of Arts and Crafts. He will transform a course for advertising draftsmanship into a complete training program in design and typography, and will be later seen as the progenitor of the Swiss School of graphic design.

Austrian-born architect, furniture designer, and set designer Joseph Urban (1872-1933) directs stage design for the Metropolitan Opera in New York, and will continue there until his death. He will also design a number of films and Broadway productions.

Richard F. Bach, hired to head the "industrial art" department at the Metropolitan Museum of Art, appeals to the annual conference of the American Federation of Art to establish "industrial arts" schools comparable to those in Europe, and to improve the artistic quality of all mass-produced household items.

Louis Cartier designs the *Tank* watch for members of the U.S. tank corps in gratitude for its defense of France during the war. The design will become very popular in the U.S.

George Hughes's company (see 1910) merges with Hotpoint (see 1905) and the heating device section of General Electric Company to form the Edison Electric Appliance Company, and to produce *Hotpoint* brand name products. Hughes is its first president. See 1922.

The *Raggedy Ann* doll, patented last year by political cartoonist John B. Gruelle, is introduced. Gruelle found a hand-made doll in his attic that he used in storytelling to entertain his dying daughter. He then published these stories as *Raggedy Ann* books. The doll will eventually grow into a $20 million per year business.

02-26. Young Dutch cabinetmaker Gerrit Thomas Rietveld (1888-1964) designs a wooden chair based on *De Stijl* principles, at the request of a client requesting chairs similar to "plank" chairs designed by Frank Lloyd Wright from 1898 to 1902 (see 1904). This chair will later be painted and titled the *Red/Blue Chair,* and though never produced, will become emblematic of the *de Stijl* movement (see 1917) and a classic of modern design.

02-27

02-27. Kelvinator introduces the first successful domestic refrigerator to enter full-scale production to the market (see 1916), which it also markets under the Leonard brand name. It uses a compression system with sulfur dioxide refrigerant developed earlier (see 1914). By 1923, there will be 56 companies in the business, but the toxic gases used will become a consumer hazard.

02-28. The U.S. government promotes air transportation by using Army planes for new scheduled airmail service between New York and Washington. The planes used are the only U.S. military aircraft still in full production, the Curtiss JN-4 "Jenny," a training plane made by Glenn H. Curtiss.

O2-29

02-30

1919

The new post-war Weimar Republic is created in Germany, which, like the Communist regime in Russia (see 1917), believes that American *Fordism* (see 1913) and *Taylorism* (see 1911) can provide a new economic base along with an agenda for social change. Reform in art education is seen as vital, because designers are needed to design high quality export items needed for economic recovery. As part of this agenda, *Das Staatliches Bauhaus* school is established in Weimar, the seat of the new government. The school's purposes are 1) to train craftsmen, painters, and sculptors of the future to combine their skills in cooperative projects, 2) to elevate the status of crafts to that of the fine arts, and 3) to establish contact with leaders in related design movements, such as *De Stijl, and Constructivism* (see 1917). The *Bauhaus* is located at the site of two pre-war schools; *Weimar Kunstgewerbeschule*, under Henry Van der Velde (see 1902), and the *Weimar Academy of Fine Art*, founded 1860. Walter Gropius accepts the position as the new director only on the condition that he is allowed to carry forward the ideas of Van der Velde.

A Constructivist exhibition is held in Moscow. El Lissitzky (Eleazar Markovich Lissitzky, 1890-1941), designs an abstract political poster promoting the Russian Revolution, *Beat the Whites With the Red Wedge*. Ukrainian Vasili Ermilov (1894-1968) uses the style and ideology of Constructivism to create posters and murals. He paints the exterior of trains with images in order to bring messages of the Revolution to outlying areas of Russia

The book *More Beautiful Things for Everyday Life*, by Gregor Paulsson, is published in Sweden. Paulsson is director of the Swedish Design Society, originally named Svenska Slöjdföreningen when founded in 1845. The more organic Swedish design philosophy will develop differently from the severe, geometrical German style. See 1916.

Johnson & Johnson Company begins to market a sterilized gauze bandage pad already affixed to adhesive tape, devised by employee Earle Dickson at home. The timesaving device will be pre-cut in quantity by 1924, and become known as *Band-Aids.* At the time of his retirement in 1957, Dickson will be a vice president of the company. See 1930.

02-29. Hobart Manufacturing Company introduces a domestic version of its commercial mixer unit used in bakeries. This mixer is named the Model H *KitchenAid,* after a housewife-tester said "I don't care what you call it, but I know it's the best kitchen aid I have ever had!" The 65-pound new electric "food preparer" is mounted on its own stand and sold door-to-door for the luxury price of $189. It will be made until 1927. See 1937.

02-30. One in ten Americans now owns a telephone. The Bell Telephone Company finally adds a finger dial to this "candlestick" style desk telephone that was introduced in 1916.

02-31. French auto firm Citroën introduces its first car, this Type A, designed by Jules Saloman. It will become the first popular mass-produced car in Europe.

02-31

Chapter 3

1920-1929 Industrial Design Practice

In postwar America, economic recovery begins as industry converts to peacetime production. Industry struggles to meet new consumer demand for products, which is stronger than ever and precipitates an economic boom.

For the first time in history, more Americans live in urbanized areas than on farms. Commercial radio reaches a vast new audience. Car production nears an incredible four million in 1923, half of which are Ford Model Ts. New highways are built and commercial aviation begins in 1925, with the Ford Tri-motor carrying passengers. The demand for home furnishings, appliances, and other consumer goods seems infinite.

The Bauhaus is the German center of design education and theory. The first public Weimar exhibition of Bauhaus work in 1923 premiers some of its best commercially produced work, but severe German inflation severely threatens survival of the school. In Sweden, design follows a softer, more organic style known as *Svenska* Form. In Russia, *Social Realism* gradually supercedes *Constructivism*.

International exhibitions of household appliances begin in France in 1923, as does the Italian Biennale, a celebration of design and architecture. The style known as *Art Deco* debuts, taking its name from the 1925 *Exposition des Arts Décoratifs et Industriels* in Paris. Design exhibitions in Germany, at Macy's in New York in 1927, and in Spain at the 1929 International Exposition in Barcelona, heavily promote new international "modern" styles in home furnishings. But while Europe sets the standards in international styles, its markets are limited, and production capabilities suffer from post-war recovery problems. The major mass market for manufactured goods is in America.

U.S. market research, a new business tool, discovers a large gap between what manufacturers make and what consumers want. Products must offer more than functionality, and the importance of visual appearance is now obvious to U.S. industry. Manufacturers begin to seek help to create more "modern looking" products. However, industry has little artistic or design talent within its corporate engineering-rich organizations.

As early as 1920, custom auto coach builders establish independent design firms and use drawings to illustrate new styling concepts to auto manufacturers. In about 1923, many design professionals in advertising, graphics, or the theater begin to recognize the growing demand for industrial product design, and respond by opening new offices or including "industrial design" in their practices. By the end of the decade, dozens of practitioners work in this new field.

In 1927, Harley Earl establishes a corporate "Art and Color" Section at General Motors, ending the market's domination by Ford's Model T. By 1929, numerous manufacturers and advertisers have also established internal industrial design groups. Style sells.

Formal design institutions are formed as well. In 1928 the American Union of Decorative Artists and Craftsmen is organized, and its founders include many of the leading new industrial design practitioners. In 1929 the Museum of Modern Art is chartered in New York, and the Metropolitan Museum of Art mounts a "Modernism" exhibition.

By the end of the decade, the new profession of industrial design emerges because of the forceful U.S. middle-class consumer market demand for new international "modern" designs. America's efficient industrial power stands ready to provide many new and fashionable products. Then the stock market crashes.

03-01

03-02

1920

An avant-garde art and design school in Moscow, VKhUTEMAS (an abbreviation of the Russian name for the Higher State Artistic and Technical Workshops), is founded. In contrast to the functions of earlier Russian art academies, this new school aims to respond to the actual psychological and physical needs of post-Revolutionary Russia in order to train artists for industry, and assist the country's cultural transformation. The school is founded on the basis of a "Production Manifesto" published by Aleksandr Rodchenko and his wife Varvarya Stepanova. The school will play a similar design role as the Bauhaus in Germany (see 1919), will reconstruct the porcelain industry as the State Porcelain Manufacturer in Lomonosov, and will industrialize the textile industry. It will become known as the VKhUTEIN in 1928, and will be dissolved in 1930.

A distinctive, narrow-necked ketchup bottle with octagonal cross-section appears in an Owens Bottle Company catalogue. Pittsburgh's H.J. Heinz Company will adopt the bottle's shape to differentiate its brand; it will become the "ketchup bottle" of the 20th century.

Philo T. Farnsworth, a 14-year-old high school student, invents electronic television. His classroom blackboard diagrams for his science teacher will later enable him to win a patent battle with RCA and, long after his death, finally be recognized as the "Father of Television." See 1939.

Lincoln Logs, a construction toy, is patented by John Lloyd Wright, the son of Frank Lloyd Wright. Inspired by traditional Japanese wooden structural earthquake-resistant joints he sees while in Japan with his father, at work designing the Imperial Hotel in Tokyo, John creates the toy system.

The Parker *Duofold* pen, nicknamed "Big Red," is introduced to the market. The pen is a refinement of the *Lucky Curve* ink feed system invented by George M. Parker, founder of the Parker Pen Company in Janesville, Wisconsin, in 1888. The system includes a feed tube that channels ink back into the ink reservoir when the pen lies on its side, thus preventing accidental blotting and leaking. "Big Red" will be nostalgically relaunched in 1994.

03-01. In Russia, artist Vladimir Tatlin creates a sculpture for the Third International celebration of the revolution. The sculpture will become a symbol of the *Constructivist* movement (see 1917). Painter Wassily Kandinsky writes the program for the Moscow Institute of Art and Culture (INKhUK).

03-02. The first radio station broadcast is made from this Westinghouse station KDKA in Pittsburgh, Pennsylvania, directed by Frank Conrad. Announcer L. H. Rosenberg tells 2,000 listeners of Warren G. Harding's election as President. By 1928 there will be 700 radio stations broadcasting.

03-03. Adolf Hitler, seventh member of the National Socialist German Workers' (Nazi) Party in Germany, designs the Party's new flag which features a swastika, or hooked cross (hakenkreuz). This symbol has been used since ancient times, but now is used for more sinister purposes. The new symbol will appear on armbands and military uniforms, and will become a powerful propaganda image for the Third Reich.

03-03

03-04

03-05

1921

The term "robot" is first used in *R.U.R.* (for Rossum's Universal Robots), a play in Prague by Czech journalist-playwright Karel Capék. Reflecting its universal meanings, "robot" stems from the Polish word for menial worker or drudge and from the Russian verb for "work".

Englishman William John Blenko (1854-1933) founds the Eureka Art Glass Company in Milton, West Virginia. He has discovered the secret of producing a brilliant "Ruby" colored glass, a highly fashionable color that will insure the company's success. His descendants will rename the company the Blenko Glass Company, famous throughout the century for creative and colorful decorative glassware. See 1941.

Giovanni Alessi founds Alessi, a company specializing in metal plate turning and a foundry for nickel, silver, and brass in Omegna, Italy. After 1927, Alessi will concentrate on manufacturing metal household items, and will become a pioneering patron of sophisticated design. See 1979, 1986, 1990.

Gabrielle "Coco" Chanel (1883-1971) introduces *Chanel No. 5* perfume in Paris. Its lack of traditional floral scent will cause it to become the world's leading perfume. Its exquisite glass bottle, the fifth of a number designed, will be the forerunner of all cosmetic packaging, and a successful classic for more than 80 years. Chanel opened her first shop in 1909, and later pioneered sportswear design for women by adapting the style and fine tailoring of men's clothing for women's apparel.

Max Braun establishes a machine company in Frankfurt, Germany to produce a device for fastening drive belts for machinery. His company will soon produce radios and phonographs, and, after 1945, will diversify into appliances. See 1955.

American architectural critic Lewis Mumford defines style as "the reasoned expression, in some particular work, of the complex of social and technological experience that grows out of a community's life." He added, "How is it that the modern style has been so slow to realize itself—is still so timid, so partial, so inadequate?"

The recent development of small fractional horsepower electric motors makes possible new appliances for kitchen use and personal care. Hand-held hair dryers, such as the *Race* by the Racine Universal Motor Company and the *Cyclone* by the Hamilton Beach Manufacturing Company, excite women of all ages.

Radio Corporation of America (RCA) begins selling mass-produced radio transmitting and receiving equipment. By 1927 there will be six million radio sets in use in the U.S.

The Tan-Sad Works of Birmingham, United Kingdom, markets one of the earliest chairs for office workers with ergonomic features—adjustable tubular steel legs and a backrest. It claims to reduce worker fatigue and increase productivity.

03-04. In Dayton, Ohio, General Motors's Frigidaire Division (see 1916) introduces its first electric refrigerator. The company's name inspires the word "fridge" as a common term that replaces "icebox." This year, 5000 refrigerators were produced in the U.S.

03-05. Allgemeine Elektricitäts-Gesellschaft (AEG), Berlin, introduces this electric flatiron with a cantilevered handle.

03-06. Black & Decker initiates advertisements for its pistol-grip electric drill (see 1917) in the *Saturday Evening Post* that stress the "manliness" of holding a pistol grip in one's hand.

03-06

03-07

03-08

1922

Willis H. Carrier, using dielene instead of the previously used ammonia, which is toxic if leaked into the air, creates the first safe air-conditioner to be used in public buildings. Carrier sells his invention to movie theatre operators, where the public first enjoys air-conditioned comfort.

Ralph W. Samuelson, 18, of Minnesota, develops water skis (two planks of wood), and related equipment such as a pull rope, a holding ring, and a ramp greased with pork fat. During the next 15 years, he will perform at water shows while being towed by a plane at 80 mph. An earlier device in 1913 was the *Aqua-Plane*, a wooden slab towed behind a motorboat. The first patented device, in 1925 by Fred Waller, will be the *Dolphin Aqua-skis,* which use two eight-foot mahogany skis towed by each tip tied to the back of a boat. The first fiberglass skis will appear in the 1960s.

Gordon Russell (1892-1980) writes a pamphlet, *Honesty and the Crafts*. In 1927 he will found Russell Workshops Ltd., in Chipping Campden in the English Cotswolds, to manufacture his own furniture designs that are influenced by the Arts and Crafts Movement. In 1929 he will form Gordon Russell Ltd., and open a showroom in London.

In Lancaster, Pennsylvania, R.E. DeWalt develops the concept of a circular saw that hangs from an arm that can be raised and lowered, as well as moved back and forth across the work. It will be called the "radial arm" saw. Black & Decker will acquire DeWalt, Inc. in 1960.

Irish interior, furniture, and architectural designer Eileen Gray opens a Paris gallery to show her furniture, lamps, carpets, and lacquer work. One of her first designs is the "Le Destin" folding screen of lacquered wood and aluminum.

03-07. Architectural renderer Hugh Ferris (1889-1962) completes *Study for the Maximum Mass Permitted by the 1916 Zoning Law, Stages 1-4* (Illustration shows Stage 2) which define skyscraper limits for architects and which will impact all new skyscraper construction and style in New York City.

03-08 . Until now, all electric ranges are in black or brown, similar to this current *Hotpoint* by the Edison Electric Appliance Company (see 1918). This year, the company receives an order from a Raleigh, North Carolina, utility executive for a white porcelain stove. Rather than admit that such a stove is not being produced, Edison quotes a high price to discourage him. Upon receiving an order anyway, they have to develop a new annealing process to apply the porcelain enamel to metal. Next year's *Hotpoint* models will include white ranges with nickel trim on the door, and white will become the industry standard. In 1931, the Edison Electric Appliance Company will become the Edison General Electric Company, and in 1934, General Electric and Hotpoint brand production will be integrated.

03-09 . Paul Jaray designs this full-scale car based on his aerodynamic models. Rudolph Ley AG, an Armstadt, Germany, automaker since 1906, builds the car. Called the Ley T6, it has a drag coefficient of 0.28 and will influence the 1933 Tatra 77, also by Jaray.

03-09

03-11

03-12

1923

The first *Biennale*, an exhibition of design and architecture, conceived by Guido Marangoni to display local work, is held in Monza, Italy. Designs include modern ceramics by Gio Ponti. The exhibition will move to Milan and become the *Triennale* in 1933, an international showcase of modern design.

The first international exhibition of household appliances in Paris, initiated by chemical engineer Jules-Louis Breton (1872-1940), attracts 100,000 visitors. In 1926 it will move to the Grand Palais, be retitled "Salon des Arts Ménages" (Salon of Domestic Science), and continue until 1939.

In the U.S., Joseph Sinel writes *A Book of American Trademarks and Devices,* which presents basic rules of corporate identity, and is illustrated with more than 300 trademarks he has designed.

In Akron, Ohio, the B.F. Goodrich Company introduces *Mystik Boots* with an innovative closure device. Company president Bertram G. Work thinks the name lacks "zip," and changes it to *Zipper Boot*, with the advertising slogan, "You just "zip 'er up or zip 'er down." Customers ignore the boot part and call the closure device a "Zipper," which Goodrich will duly register and promote. The closure device has been on the market for some time, but was called a "Hookless Fastener" (see 1914). In 1928 the trade name of the fastener will be changed to *Talon,* and in 1937 the Hookless Fastener Company will change its name to Talon, Inc., which in 1968 will be acquired by Textron.

Beatrice Alexander Behrman, known later as "Madame Alexander," founds the Alexander Doll Company. The company makes dolls with molded faces that become popular because German dolls have not been made since the First World War halted production nearly ten years ago. Her business will skyrocket in 1934 with dolls patterned after the Dionne quintuplets. She will die in 1990 at age 95, but her business will continue.

03-10

03-10. Wilhelm Wagenfeld and Karl Jacob Jucker (1902-1997), students at the Bauhaus, design this table lamp (MT 8) with a hemispherical glass globe.

03-11. The first full-scale exhibition of Bauhaus (see 1919) work is held in August at the school. It includes an experimental house "Haus am Horn" designed by George Muche with this kitchen designed by Benita Otte and Ernst Gebhardt and a ladies bedroom and dressing table designed by Marcel Breuer. Johannes Itten resigns and László Moholy-Nagy begins teaching as head of the metal workshop.

03-12 . On September 1, opening day ceremonies at the Imperial Hotel in Tokyo, Japan, are interrupted by the Great Kanto earthquake and hurricane winds that lay waste to the city. The hotel not only survives with just minor damage, but also becomes a home for refugees. Frank Lloyd Wright had been working since 1916 on the hotel project which was plagued with problems of fire in 1920, and then burned to the ground in 1922. Now Wright's design is vindicated and his reputation assured. Wright also designed oak chairs with hexagonal backrests for the hotel in 1920.

03-13

03-14

1924

Danish furniture designer Kaare Klint (1888-1954) joins the faculty at the Copenhagen Academy, and teaches furniture design and craftsmanship. He will later teach architecture, and influence the development of "Danish Modern" furniture design.

The Metropolitan Museum of Art holds another of its industrial arts "quantity production" exhibition series begun in 1917, curated by Richard F. Bach, who stipulates that objects entered must be based on museum collections. In fact, this means that the designs are to be custom designs, not manufactured in quantity.

The DeVilbiss Manufacturing Company of Toledo, Ohio, hires Frederic Vuillemenot, a graduate of the Ecole des Arts Décoratifs in Paris, as chief designer for the company's elegant perfume atomizers and accessories.

Artist John Vassos receives his first industrial design assignment from cosmetic firm Armand Products to create a face lotion bottle with a new plastic screw cap. The bottle becomes immediately popular as a liquor flask during Prohibition.

The metal workshops at the Bauhaus (1919), directed by László Mohóly-Nagy, produce outstanding designs by students. These soon include a chrome-plated teapot by Naum Slutzky, sugar and creamer by Marianne Brandt (1893-1983) (see illustration), and experimental tubular steel chairs by Marcel Breuer and Mart Stam (see 1927).

William Petersen, a blacksmith in DeWitt, Nebraska, patents the *Vice-Grip* pliers with a toggle-locking lever that he invented in 1920. He will manufacture them and, as a traveling salesman, sell them from the trunk of his car. After World War II, sales will take off; by 1997 Petersen Manufacturing, run by William's grandson Allen, will sell millions per year.

Kimberly & Clark Paper Company introduces *Celluwipes,* a new product made from cellucotton (a cotton-like cellulose material), as a somewhat costly cosmetics remover for well-to-do women. A pop-up-box by Chicago inventor Andrew Olsen was added in 1921, but *Celluwipes* will not do too well on the market until 1930, when their frequent use as disposable handkerchiefs is discovered and promoted as *Kleenex.*

03-13. The first Chrysler car, developed by Walter P. Chrysler (1875-1940) and designed by engineers Carl Breer (1883-1970), Fred Zeder, and Owen Skelton, is introduced by the Maxwell Motor Company in a hotel lobby because it is denied space at the Detroit Auto Show. Chrysler had been executive vice president of General Motors, but left in 1920 because of disagreement with the direction of the company; he became president of Maxwell last year. Maxwell's innovative new car features improved 4-wheel hydraulic brakes, and will prove to be the last successful start-up of a new car company in the U.S., as next year, Maxwell will be named the Chrysler Corporation. By 1929 it will become one of the "Big Three," joining General Motors and Ford.

03-14. Suger and creamer by Bauhaus student Marianne Brandt.

03-15. The Electrolux Model 5, a cylinder or "tank" vacuum cleaner, is introduced in the U.S. The Atkibolaget Electrolux Company in Sweden created the style, then named Model III, in 1915. It uses a "clean air " system that collects the dirt in a bag before the air is drawn into the motor by the fan.

03-15

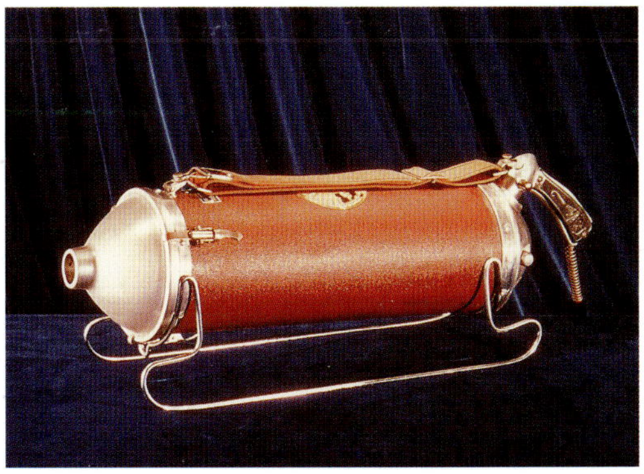

03-16

03-17

1925

"Art Deco" style reaches its zenith with the highly successful *Exposition Internationale des Arts Décoratifs et Industriels- Modernes* in Paris. U.S. manufacturers do not participate in the Exposition, since President Coolidge told organizers that there was no American design modern enough to warrant representation. But the U.S. sends a commission, which alerts industry to the necessity of meeting foreign design competition. The commission recognizes that the modern movement is international in nature and will soon reach the U.S. If the U.S. does not use the movement to enhance modern living conditions and tastes, it will be dependent upon foreign talent, designers, and manufacturers. Soon, reviews of the exhibition will sweep away much of the American resistance to Art Deco, and popularize the terms "Art Deco," "Art Moderne," and "Modernistic".

Cranbrook Academy of Art is founded on the Bloomfield Hills, Michigan, estate of George Gough Booth, a Canadian-born newspaper publisher, and his wife Ellen Scripps Booth. They engage Eliel Saarinen, Finland's leading architect, as the Academy's designer, president, and guiding force.

The Leica 35 mm camera becomes available to the public as Model A. Designed by Oskar Barnak (1879-1936), a microscope designer who worked for Ernest Leica at the Leitz Optical Works in Wetzler, Germany, and applied the principle of 35 mm motion-picture film to still photography.

Chromium, discovered in 1798, used for projectile covering during World War I, but not commercially available until this year, is adopted as a brilliant and corrosion-resistant finish for tubular steel. See 1927.

03-16. The *Exposition* includes *L'Esprit Nouveau,* an exhibit that premieres a housing unit commissioned by the magazine of the same name, designed by Le Corbusier and his cousin Pierre Jeanneret (1896-1967). Ignored by the press, the exhibit is characterized by an essentially unfinished interior outfitted with commercial furniture, metal filing cabinets, bistro wine glasses, laboratory flasks used as vases, and industrial equipment in the "décor." The room expresses a complete rejection of traditional decorative arts, but within five years its emphasis on making the home a more efficient place will greatly influence the direction of the modern movement.

03-17. The Bauhaus (see 1919) moves from Weimar to Dessau, shifting its philosophy from rarefied tastes and exoticism to science and technology. Dessau is an industrial city, and Walter Gropius, influenced by Ford (see 1913) and Taylor (see 1911), advocates the standardization of precision parts, the division of labor, and the use of capital-intensive, labor-saving, special-purpose machinery. The school will in 1926 become the *Hochschule für Gestaltung* (Institute of Design) as it opens its new building designed by Gropius. Faculty members this year include Josef Albers, Herbert Bayer, Marcel Breuer, and Gunta Stözl.

03-18. The first commercially successful all-metal airplane, the Ford Tri-motor 3AT (known as the "Tin Goose"), enters air service between Detroit and Chicago for Stout Airlines. It was made of Duralumin in 1922 by William Bushnell Stout, who sold his company to Ford, became vice president and general manager of its Stout Metal Airplane Division, and founded Stout Air Services. See 1934.

03-18

03-19

03-20

03-21

1926

In England, John Logie Baird (1888-1946) successfully demonstrates his "televisor," a mechanical system of television using spinning disks developed in 1884 by German engineer Dr. Paul Nipkow. The flickering neon-pink image is transmitted several yards. Baird invented the system in 1924 and, by 1928, will be able to achieve crude color television transmissions.

The Hoover Company introduces its Model 700 upright vacuum cleaner. The machine's innovative rotating cylindrical agitator with beater bars and bristles to loosen dirt inspires the company's slogan, "It beats, as it sweeps, as it cleans" that will become Hoover's key competitive sales weapon.

Porter-Cable Division of Rockwell Manufacturing Company in Syracuse, New York, enters the portable power tool market with its "Take-About" portable belt sander, the first such tool in the industry, designed by Art Emmons. The company was founded in 1906 to make pencil sharpeners.

Richard G. Drew (1886-1956) of the Minnesota Mining and Manufacturing Company, founded in 1902, devises "masking tape" a solution that results in clean two-tone auto painting. Drew had tried to save money by putting adhesive only on the outer edges of the tape, but a body shop foreman found that the tape did not stay in place. The foreman complained to the salesman. "Take this tape back to those stingy Scotch bosses, and tell them to put adhesive all over." Workers nickname it "Scotch Tape," and the name will stick, too. Soon, a transparent version (also perfected by Drew) will be produced (see 1931).

03-19. The Waters-Genter Company, formed in 1921 in Minnesota to produce the automatic toaster patented by Charles Strite in 1919, introduces *Toastmaster,* the first consumer pop-up toaster. The Model 1-A-1 will become popular, despite its luxury cost. Toasters will assume an important position in the electric household and will offer a great variety in technology and design in the next decade. The company will be acquired later this year by McGraw Electric Company, and in the future will be known as McGraw-Edison, then as Toastmaster. By 1990, the brand will be owned by Salton Corporation.

03-20. With this new model, Chevrolet, a Division of General Motors, challenges Ford with bold new styling and colors, forcing a change in the Model T, which Ford now offers in dark green and dark maroon in addition to its standard black. By 1927, General Motors will capture 45 percent of the U.S. market.

03-21. Car designer John Tjaarda begins work on this aerodynamic vehicle that he names the *Sterkenberg Series* after his family ancestral estate in Holland. He will refine the concept by 1930, and it will become the basis for his influential "dream" car for Briggs and Ford (see 1933).

03-22. Robert Hutchings Goddard (1882-1945) launches the world's first liquid-fueled rocket on March 16, for a 2.5 second, 40-foot flight near Auburn, Massachusetts. In 1921 Goddard had explained in a *Scientific American* article exactly how a rocket could be built to reach the moon. Ignored by the U.S. government, his patents will enable Germany to build guided missiles.

03-22

03-23

03-24

1927

A *Deutscher Werkbund* (see 1907) exhibition with the theme of "The Home" presents houses for modern living in Stuttgart under the direction of Mies van der Rohe. Known as the *Weissenhof Siedlung* (Weissenhof Estate), the exhibition includes tubular steel chairs by Mart Stam, Marcel Breuer (see 1924), Mies van der Rohe, and other designs developed for the public housing boom in Weimar Germany.

Macy's of New York holds a weeklong *Exposition of Art in Trade*, "contrived in modern spirit", and filled with "modern products" viewed by 40,000 visitors at the store. A week later, a *Machine-Age Exposition* opens in New York City for two weeks, organized by Jane Heap, co-editor of *The Little Review,* an avante garde art publication. The exhibit promotes machines as art, and the idea of engineers and artists working together.

Numerous prominent U.S. designers initiate practice in the new field of industrial design. Designer Walter Dorwin Teague adds the term "industrial design" to his letterhead, and contracts his first retainer with Eastman Kodak at $15,000 per year. Theatrical set designer Norman Bel Geddes begins designing Graham-Paige cars. Package designer Egmont Arens works on the design and manufacture of a line of U.S. lamps. George Sakier serves as design director of the American Radiator and Standard Sanitary Corporation.

At General Motors, Harley Earl establishes the Art and Color Section, which will be renamed the Styling Section starting in 1937, and introduces the use of full-scale clay models as the best method to visualize and develop a car's form. Within a year he will recruit fifty people, many of whom will later direct styling departments at all major automakers for decades, and thus will prolong Earl's influence on auto design.

03-25

03-23. At the *Weissenhof Seidlung*, architect Marcel Breuer exhibits his tubular steel chair. Inspired by the lightweight construction of his bicycle, Breuer began developing this revolutionary concept two years earlier. The chair, named Model B3, is manufactured by Standard-Möbel of Berlin. Known as the *Wassily* chair, after Wassily Kandinsky, Breuer's mentor at the Bauhaus, the chair will become a classic of modern design and will be produced under license by Gebrüder Thonet in 1928.

03-24. This cantilevered, curved tubular chrome-plated steel chair is shown at the *Weissenhof Siedlung*. Designed in 1926 by Mies van der Rohe, it is produced by Joseph Muller to furnish an apartment building designed by Mies for the same exhibition. Future reproductions, known as MR or *Weissenhof* chairs, will be made by Thonet beginning in 1931, and by Knoll Associates after 1964.

03-25. General Electric (GE) introduces this "Monitor Top" refrigerator, so named because the shape of its exposed refrigeration unit on top of the cabinet reminds people of the cylindrical turret of the Civil War gunship. Designed by Chief Engineer Christian Steenstrup (1873-1955), the refrigerator is the first to use an all-steel cabinet and sells for $525. The Monitor will make GE the industry leader and will become one of the company's most successful products.

03-27

03-28

03-29

1928

Under increasing criticism from the Socialist Democratic government of Dessau, Walter Gropius resigns from the Bauhaus. Gropius appoints Hannes Meyer as his successor, and Josef Albers succeeds Breuer as head of the furniture workshop. See 1925.

Macy's opens its second (see 1927) *International Exposition of Art in Industry*, providing an opportunity for American designers to show their product designs in the new modern styles. Exhibitors include Kem Weber, Peter Müller-Munk, Josef Hoffmann, Bruno Paul, William Lescaze, and Eugene Schoen.

The American Union of Decorative Artists and Craftsmen (AUDAC) is founded to define the new profession of industrial design. Founders include Donald Deskey, Paul T. Frankl, and Lucian Bernhard. Later members will include Frank Lloyd Wright, Walter Dorwin Teague, Norman Bel Geddes, Russel Wright, and Hugh Ferris. AUDAC will organize two exhibitions in 1930 and 1931.

Demonstrating the alliance of art and industry, the School of Design at The Toledo (Ohio) Museum of Art establishes specialized courses in industrial design at the request of individual local companies, such as the DeVilbiss, Owens-Illinois Glass, the Toledo Scale, and the Libbey Glass Manufacturing companies. Idene McAleese (later Ayers) is one of the instructors.

"Skyscraper" style flourishes, with a height and weight meter designed by Jo Sinel for International Ticket Scale Corporation, and a "Skyscraper" tea service, brush, and mirror designed by Louis B. Rice for Bernard Rice's Sons, Inc. *Apollo Studio* line.

The Libbey Glass Manufacturing Company introduces a *No-Nik Safedge Bell Fountain* beverage glass pattern for use in soda fountains. The beaded lip on the rim is extremely resistant to chipping, insuring its long-lasting sales appeal. The unique design will be produced for Coca-Cola for many years, and will be associated in the minds of consumers as a partner to the drink's trademark bottle shape (see 1915).

Chrysler Corporation introduces its first Plymouth automobile, and establishes an Art and Color Department of five employees, headed by industrial designer Herbert V. Henderson. In 1932, Chrysler will hire Raymond Dietrich as a consultant, who will serve as unofficial head of the Department until 1934, then officially until about 1940.

03-27. Eliel Saarinen designs these chairs for a Boy's School that opened last year at Cranbrook Academy of Art (see 1925). The chairs will still be in use in 1995.

03-28. The Leipzig firm of Körting & Matthieson produces quantities of more than 50 models of lamps designed at the neighboring Bauhaus between this year and 1932. An early example is this *Kandem* bedside table lamp by students Marianne Brandt and Hin Bredendieck, their most commercially successful design. It exemplifies the Bauhaus philosophy of making products "at a reasonable price through the utilization of all the modern economical means of standardization (industrial mass production) and wide marketing," as conceived by Walter Gropius.

03-29. Körting & Matthieson produce a similar *Kandem* writing table lamp, also designed by students Hin Bredendieck and Marianne Brandt.

03-30

03-31

03-32

1929

The stock market crash reduces Art Deco's attractiveness because the style's craftsmanship and fine materials become too expensive. The shimmering surfaces and angular shapes will continue to be used for a few years in theatres, skyscrapers, and luxury ocean liners.

Advertising pioneer Elmo Calkins establishes a styling and design department at his New York agency, appointing Egmont Arens, a New York editor of fashion magazines, as director. Calkins, who is deaf, hails design as "the silent salesman".

Business Week magazine publishes several articles on "modern art and its relation to business" that observe "radical changes in our ideas of color and design," and that the artist had become a businessman who could "stylize products."

In Pittsburgh, Pennsylvania, Westinghouse Electric and Manufacturing Company hires Donald Dohner as Director of Art. When he leaves Westinghouse in 1934, he and his staff of eight will have designed 128 products, including electric ranges, diesel-electric locomotives, water coolers, and ashtrays.

Donald F. Duncan, who acquired rights from a Filipino immigrant, Pedro Flores, markets the Yo-Yo in the U.S. The toy will be heavily promoted and become a national fad.

Errett Loban Cord (1894-1974), president of the Auburn Automobile Company in Auburn, Indiana, introduces the *Model J* Duesenberg. For the next few years, the car is designed by the best coachbuilders, including Gordon Buehrig. At a retail price of $14,000 each, 481 will be sold until 1936. Purchased by movie stars such as Clark Gable and Gary Cooper, the extravagant car inspires the slang phrase, "It's a real Doozy!" See 1936.

Sir Herbert Austin (1866-1941) of England forms the American Austin Car Company to build a tiny car designed by Alexis de Sakhnoffsky for the American market. It will become popular in the U.S. as a prop for Hollywood films and photographs. Austin will reorganize as the Bantam Automobile Company in 1936, with Thomas L. Hibbard as designer, and will originate the *Jeep* (see 1941).

03-30. Mies van der Rohe and Lilly Reich design these chairs to serve as ceremonial seats for the King and Queen of Spain at the German pavilion of the International Exposition in Barcelona. Various manufacturers will produce replicas of what will become known as the *Barcelona* chair.

03-31. Thonet introduces the Model B 32, Marcel Breuer's first cantilevered tubular steel chair. It will have the most widespread influence of any tubular steel chair. Reproduced in 1962 by Gavina, it will become known as the *Cesca,* a diminutive form of Francesca, Breuer's daughter's name.

03-32. In their goodwill flight around the world, Charles Lindbergh and his wife, Anne Morrow Lindbergh, fly this *Sirius* monoplane, designed at Lockheed Aircraft Company in Burbank, California, by chief engineer John K. Northrop, Gerald Vultee, Richard von Hake, and Allan Loughead. Allan and his brother Malcolm founded the company in 1916.

03-33. Richard Bach, Director of the Metropolitan Museum of Art, opens a "Modernist" exhibit titled *The Architect and the Industrial Arts*. Eliel Saarinen designs every element of a dining room ensemble, including this *Silhouette* silverware.

03-33

Chapter 4

1930-1939 Streamlines

The Depression slows momentum, but with impressive new public and private construction projects under President Franklin Roosevelt's New Deal, the decade soon turns more optimistic with the 1933-1934 *Century of Progress* Exposition in Chicago. The grand fair becomes a training ground for many young industrial design practitioners, and a showcase for the latest achievements in transportation and consumer products.

The first clear description and important exposure that legitimizes industrial design appears in the February, 1934, issue of *Fortune* magazine, in which an article describes the emerging profession as the "wave of the future." Business "discovers" industrial design, and manufacturers pay incredible fees to designers whose stylish designs are enormously successful in markets tired of the mechanical, awkward appearance of most consumer products of the 1920s. In product after product, sales increase by hundreds of percentage points. The new profession is "hot," and "modern" design is now expected in all products. Educational programs in Industrial Design are initiated, including some by the federal government, and manufacturers and retailers establish new internal industrial design groups to create and satisfy consumer demand.

Consumer products are now being brought to mass production reality through industrial design in the U.S.; not by academic theory or imitation of European approaches, but by direct interaction with business clients, and because of the demonstrated evidence of public mass-market acceptance. No longer looking to Europe for leadership in design and fashion, Americans now embrace their own styles that express speed, modernity, and technology.

Fascination with high-speed travel inspires "streamlined" products such as the Chrysler *Airflow* automobile and the *Zephyr* passenger train. This peculiarly American product style symbolizes a better future to a generation scrimping through the severe Depression of the early 1930s. Streamlining represents the American ideal of progress, and enhances, shapes, and rejuvenates everything from cars to pencil sharpeners.

Kitchens again bring the future into the home, as refrigerators, stoves, and sinks, which were initially free-standing monuments of engineering, evolve during this decade into the simplified white shapes integrated into the kitchen counters and cabinetry that are familiar today.

The decade will be remembered as the golden age of U.S. industrial design, because the public recognizes design as the creative force that gives physical form to the American Dream. The New York World's Fair of 1939, with its gleaming Oz-like cities in the center of bucolic suburbs and modernistic superhighways, strives to present this dream for a life 20 years hence. People enthusiastically embrace these utopian visions of the future.

The decade also proves to be one of the most productive in the research and development of new materials, including polystyrene, fiberglass, synthetic rubber, *Nylon*, melamine, polyurethane, and *Teflon*. All of these will provide new inexpensive materials for mass production, even though the war will delay their full impact and application to consumer markets.

By the end of the decade, industrial design can boast of industry acceptance, formal design organizations, and educational models. Its pioneers are now its leaders, ready to convey their hard-won knowledge and expertise to the second generation of designers. But soon, the clouds of war will push aside the realization of the future.

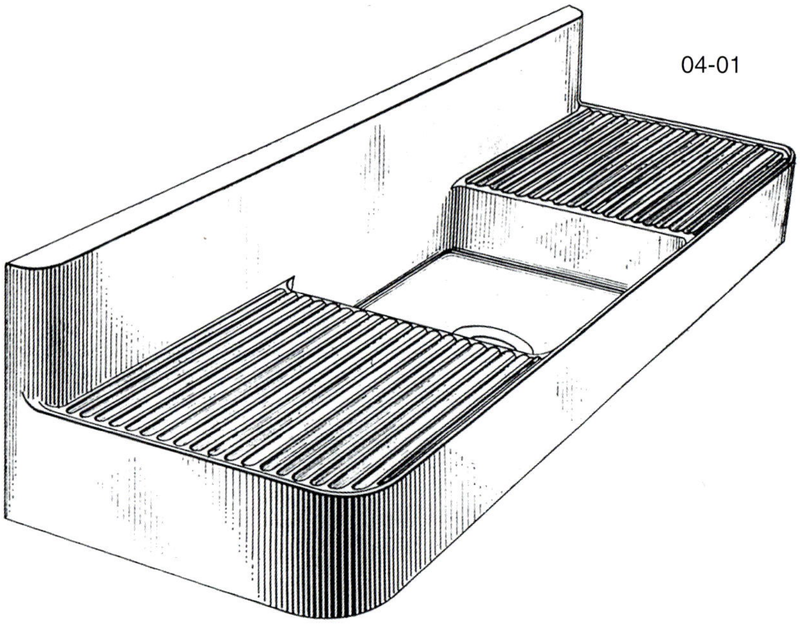

04-01

04-02

04-03

1930

Polystyrene, a plastic that is cheaper, faster molding, dimensionally stable, and with greater color clarity than previous plastics, comes into common usage. Developed by I. G. Farben in Germany, it will be produced commercially in 1935, and will become a major influence in consumer product design after the war.

William Chambers, a student at McGill University, Canada, invents acrylic, a clear thermoplastic material. It will be marketed in Britain as *Perspex.* Acrylic will appear in the U.S. with trade names of *Plexiglas* and *Lucite.* (See 1937)

Chicago Flexible Shaft Company introduces the *Mixmaster*, a versatile food-preparation appliance that will become a staple item in American kitchens. Chief Engineer Ivar Jepson will improve the Mixmaster's features, and will design hundreds of products during his more than 20-year career at the company. Chicago Flexible Shaft will be renamed Sunbeam Corporation in 1946, and its products will become classics of the American home.

Italian Alfonso Bialetti designs the *Moka Express*, a stovetop espresso coffeemaker made of cast aluminum shaped with eight flat planes in order to withstand high steam pressure. The product will not sell well until the 1950s, when Alfonso's brother, Renato, will market it, but by 2000 it will be found in 250 million households worldwide.

Johnson & Johnson Company (see 1919) invents a white waterproof cloth tape that they name *Drybak,* for use in hospitals. In 1942 it will be sold as white "Duck tape" for window blinds, and in 1970, with the color changed to gray, air-conditioner installers will use it to seal ducts. Building contractors will depend on this "duct tape," and consumers will invent a thousand other applications.

One in five Americans, or more than half of all U.S. households, owns a car. General Motors, through the talents of its new Art and Color Department under Harley Earl (see 1927), has evolved five distinct brand images. *Fortune* magazine characterizes the consumers for each: the Chevrolet, for the "hoi polloi," Pontiac, for "the poor but proud," Oldsmobile, for "the comfortable but discreet," Buick, for "the striving," and Cadillac, for "the rich."

04-01. This monel metal sink is designed and patented by Gustav B. Jensen for the firm of Byrnes, Stebbins, Parmelee & Blenko of Pittsburgh, Pennsylvania. The design was for the International Nickel Company and shown at the American Union of Decorative Artists and Craftsmen (AUDAC-) (see 1928) exhibition at the Grand Central Palace in New York.

04-02. Mies van der Rohe designs this armchair for the airy interior of a dramatic glass and steel house he has designed for the Tugendhat family in Brno (Brünn), Czechoslovakia. In 1960 a version known as the *Brno Chair*, will be reproduced by Knoll for the *Four Seasons* restaurant in New York designed by architect Philip Johnson. Another version, known as the *Tugendhat Chair,* will be reproduced by Knoll in 1964.

04-03. Miele in Germany is marketing this *Duplikat,* Europe's first electric household dishwasher, that it introduced last year. Filled by hand, the machine requires about eight minutes to complete two soap and rinse cycles.

04-04

04-05

04-06

1931

The Metropolitan Museum of Art opens *Contemporary American Industrial Art*, an exhibition now broadened to include many manufactured products.

In a shift from traditional to modern design, The Herman Miller Furniture Company of Zeeland, Michigan hires Gilbert Rohde. The company was founded in 1905 as the Star Furniture Company in Michigan, and renamed in 1923 when Herman Miller and his son-in-law, D.J. DePree, purchased controlling interest. Rohde met DePree in 1929 and prepared some designs for him. His first modern design for the company, the No. 2185 Group bedroom suite, will be produced in 1932.

Anne Swainson is named the first director of the new Bureau of Design at Montgomery Ward in Chicago. She will manage the graphic design of catalogues, packaging, and product design. Because of the great variety of products, the Bureau will become an important training ground for many designers. By 1935 Dave Chapman will head a staff of 18 at the Bureau. The Bureau will be discontinued in the late 1940s, then reestablished in 1955 by Chapman.

Schick Electric introduces the *Colonel*, the first electric shaver, priced at $25. Colonel Jacob Schick founded Schick Dry Shaver, Inc. in Stamford, Connecticut, in 1930. Renamed Schick Electric, the company will make injector razors and blades in Milford, Connecticut, through 1995. Schick Electric will move to Lancaster, Pennsylvania, in 1955, will be sold in 1957, and will go out of business in 1979.

The Toledo Scale Company debuts a colorful coin-operated skyscraper style Public Health Scale designed by Harold Van Doren and J. Gordon Rideout, who formed their partnership this year in Toledo, Ohio. In order to complement colorful shop interiors, the porcelain-enameled scale rejects the standard white finish, and is offered in several colors. This innovative scale changes the market for public weighing scales because the weight indicator is visible only to the person standing on the platform.

The Minnesota Mining and Manufacturing Company (3M) (see 1926) introduces Scotch cellophane tape, invented by employee Richard G. Drew as a waterproof packaging tape for food. The tape will become an item used in offices and households worldwide.

04-04. Hermann Gretsch, an architect working for Porzellanfabrik Arzberg in Bavaria, Germany, founded in 1888, designs this Arzberg porcelain dinnerware. The design will win gold medals at exhibitions in Milan (1936) and Paris (1937), and will be imported to the U.S. in 1950, when it will be the only affordable contemporary design fine dinnerware available.

04-05. This low leather couch framed in wood with tubular steel legs designed by Mies van der Rohe is introduced at the Berlin Building Exposition. In 1964, Knoll Associates will initiate reproductions.

04-06. This Pan-American Airlines amphibious flying boat, the S-42, enters service. It is designed and built by Russian aircraft designer Igor Sikorsky, who emigrated to the U.S. from Russia in 1923 and will invent the first successful helicopter in 1939.

04-08

04-09

1932

In January, the Dessau City Council dissolves the Bauhaus due to "Jewish-Marxist art manifestation". Mies van der Rohe moves the school to Berlin as a last ditch effort, opening the winter semester on Birkbuschstrasse in the suburb of Steglitz, but next year, the school will be closed.

The *Design for Machine* exhibit in Philadelphia features an all-aluminum breakfast room designed by Russel Wright that features lamps, tables, accessories, and flatware in aluminum. Aluminum had been used primarily for cookware. Wright is credited as the first designer to "bring aluminum out of the kitchen," and in future years designs many affordable tableware items in this material.

Little, Brown & Company publishes Horizons, a book by Norman Bel Geddes. The book promotes his many dramatic and futuristic designs, and is illustrated by his employee, C. Stowe Myers. Bel Geddes states, "In the perspective of fifty years hence, the historian will detect in the decade of 1930-1940 a period of tremendous significance...doubtless he will ponder that, in the midst of a world-wide melancholy owing to an economic depression, a new age dawned with invigorating conceptions, and the horizons lifted."

George Carwardine (1857-1948) designs the *Anglepoise* lamp for Herbert Terry & Sons, England. Articulated like a human elbow, it uses springs instead of the counterweights used in earlier lamps made and patented in France by Edouard-Wilfrid Buquet. The *Anglepoise* lamp will come to the U.S. as the *Luxo* lamp (see 1951).

04-07

Belle Kogan opens her design office in New York. Believed to be the first female industrial designer, she began her career in 1929 as a designer for the Quaker Silver Company, and will achieve prominence in the profession from 1935 on, designing houseware products.

The Greene Manufacturing Company of Racine, Wisconsin, uses a food mixer to "reduce fruits and vegetables to a fluid state." Green's new employee Stephen J. Poplawski patented the mixer in 1922 while working previously for the Arnold Electric Company. The Greene Manufacturing Company, founded in 1924, will become the John Oster Manufacturing Company, and will claim that this is the original "blender". See 1937.

There are now 19.2 million U.S. households with radios. These radios typically occupy a prominent place in the home as living room furniture. Blaupunkt, a subsidiary of the German firm Robert Bosch Corporation, markets a radio that is operated from a car battery with an aerial.

In Germany, Dr. Ferdinand Porsche develops the *Kleinauto,* or Type 32, an inexpensive rear-engine small car for the NSU Company. The concept will be refined, with the support of Adolf Hitler, into three 1936 VW 3 *Volksauto* prototypes *and* pre-production *Volkswagens.* See 1938.

04-07. This blue Art Deco "Jazz Bowl" is created by Viktor Schreckengost, Cleveland School of Art faculty member, while working part-time for the Cowan Pottery Studio.

04-08. The Museum of Modern Art presents its first architectural exhibition, the *International Exhibition of Modern Architecture* and publishes the book, *The International Style: Architecture Since 1922.* The term *International Style,* coined by curators Henry-Russell Hitchcock and Philip Johnson, will become the accepted term to describe such spare and unadorned forms.

04-09. This Marmon 16 auto is introduced, designed by Walter Dorwin Teague in collaboration with his son, Walter Dorwin Teague, Jr., and engineer Howard C. Marmon,

04-11

04-12

04-13

04-14

1933

Rockefeller Center opens in New York, dubbed "Radio City" because its space is leased to the Radio Corporation of America (RCA) and the National Broadcasting Company (NBC). Donald Deskey's designs for its Music Hall, interiors, and furnishings popularize Art Deco. The Music Hall will be completely restored in 1999.

The *Century of Progress Exposition* opens in Chicago. Running for two summers (see 1934) it is a showcase for modern design and technology. It includes the Ford Rotunda by Walter Dorwin Teague, and sculpture by Alfonso Iannelli for the Social Science Hall. The Perey Company provides turnstiles, named "passimeters," designed by John Vassos to count the crowds. The fair attracts design talent to prepare the many buildings, and many young designers start their careers by working on the Exposition's staff or as consultants.

Radio Corporation of America (RCA) establishes an internal design staff headed by John Vassos, and General Electric establishes an internal industrial design group in Bridgeport, Connecticut, headed by Raymond Patten. See 1948, 1985.

Zippo Manufacturing Company, Bradford, Pennsylvania, introduces the Zippo lighter, designed and manufactured by George Grant Blaisdell (1895-1978) who had purchased rights to an Austrian lighter and redesigned it. The wind-proof Zippo will become a legendary symbol of masculinity, and will still be manufactured at the end of the century, with sales of nearly 400 million units.

British chemists R.O. Gibson and E.W. Faewcett of Imperial Chemical Industries create polyethylene, the first man-made polymer. Although it will appear in some products in 1942, it will be more widely used in the early 1950s. See 1947.

The Nazi government produces a plastic molded radio cabinet designed by Walter Maria Kersting. Hagenuh makes the VE 301 *Volksempfänger* (People's Radio) in Kiel, Germany. By 1939, 12.5 million will have been sold.

The Gestetner Model 66 office duplicator is introduced. Raymond Loewy's first major design, it is considered by many to be the first "industrial designed" manufactured product. See 1908.

04-11. General Electric introduces this new "flat-top" refrigerator (illustrated) and a modified version of its *Monitor-Top* (see 1927); both designed by Henry Dreyfuss. The "flat-top," introduced under the *Hotpoint* name, places the refrigeration unit inside the cabinet, rather than above.

04-12. This sterling silver flatware is designed and produced by Russel Wright. The Museum of Modern Art will later produce replicas.

04-13. Buckminster Fuller (standing in photo) demonstrates his teardrop-shaped *Dymaxion Transport Unit* at the *Century of Progress.* Designed by Fuller and Starling Burgess, the car is propelled by two front wheels, with a third wheel in the rear that the driver controls with the steering wheel. Three vehicles have been built, but a fatal accident in 1935 will doom the project.

04-14. At the *Century of Progress,* Ford displays this *Briggs Dream Car*, a rear-engine car designed by John Tjaarda of Briggs Manufacturing Company, Detroit's largest car body producer. Based on Tjaarda's *Sterkenberg* designs (see 1926), the design will become the prototype of the Lincoln *Zephyr* (see 1936). Ford also begins annual styling changes now.

04-15

04-16

04-18

04-17

1934

George Nelson, writing anonymously, writes "Both Fish and Fowl," and article in *Fortune* magazine. It describes the work of numerous leading designers, and raises their prestige throughout industry. The article, which promises spectacular commercial success for designer-enhanced products, will legitimize the new industrial design profession.

The *Art and Industry* exhibition at Rockefeller Center is organized by designers and sponsored by the National Alliance for Art and Industry. Featuring 1000 products by 100 designers, the exhibition illustrates, "what designers are doing....to help the engineer sell his mechanical devices."

The Metropolitan Museum of Art's exhibition, *Contemporary American Industrial Art*, features a full-scale representation of Raymond Loewy's design office, and work by many other designers.

Carl Sundberg and Montgomery Ferar establish a design office in Detroit after both are fired from General Motors Styling Section on the same day.

In November, Carnegie Institute of Technology (after 1967, Carnegie Mellon University) in Pittsburgh, Pennsylvania, initiates the first undergraduate degree in industrial design, with faculty members Alexander Kostellow, Donald Dohner, Robert Lepper, Wilfred Readio, Russell Hyde, and Frederick Clayter. See 1936.

The *International* table radio, designed by John Vassos, is RCA's first made in plastic, and replaces the common cabinet radios Vassos termed "wooden tombstones." Vassos also designs the first console radio without obvious "legs."

DuPont chemist Wallace Hume Carothers invents *Nylon*, an artificial silk. It will be patented in 1938, and the name given to a family of polyamide resins which will follow. Early products will be toothbrushes, such as the 1938 Dr. West's Miracle Tuft, and nylon stockings in 1940.

04-19

04-15. The *Machine Art* exhibition is held at the Museum of Modern Art. Architect Philip Johnson (b. 1906) curates the exhibition, which celebrates "ideal shapes" of machine-made products. Alfred H. Barr, Jr., director of the Museum, writes, "Machines are, visually speaking, practical applications of geometry." The Museum initiates its permanent design collection with objects selected from the *Machine Art* exhibition.

04-16. The Wooster (Ohio) Rubber Company produces the first "Rubbermaid" product, this dustpan in variegated colors of molded rubber patented by James R. Caldwell and his wife Madeleine. The new company will be incorporated in 1938, and will be re-named *Rubbermaid* in 1957.

04-17. The Burlington Route's aluminum, diesel-powered, streamlined *Silver Streak Zephyr* trains, designed by Albert Dean of the E.G. Budd Manufacturing Company, arrives at the *Century of Progress Exposition* (see 1933) after a record-breaking trip from Denver. The entire train weighs the same as a single conventional train car and travels at 112 m.p.h. Chicagoan Helen Hughes Dulany designed the stainless steel dishes used in the dining cars of the Burlington line.

04-18. Chrysler introduces this *Airflow* auto, designed by chief engineer Carl Breer (see 1924), Fred Zeder, and Owen Skelton. It will fail disastrously in the market but innovative features will advance automotive engineering, and its form will significantly influence design. Photo shows DeSoto version.

04-19. Eastman Kodak Company introduces the *Baby Brownie* camera, designed by Walter Dorwin Teague. Four million will be sold at $1.00 until 1941.

04-20

04-21

1935

The Works Progress Administration (WPA) establishes the Design Laboratory, a design school for industry. Gilbert Rohde is appointed director with Walter Dorwin Teague, Henry Dreyfuss, and others on the Board. It will be renamed the Laboratory School of Industrial Design, and will survive until 1940.

Donald Dohner, a consultant and lecturer at Pratt since 1932, initiates a three-year certificate program in industrial design at Pratt Institute in Brooklyn, New York. See 1943.

Harold Van Doren and J. Gordon Rideout patent streamlined children's toy vehicles for the American National Company of Toledo including the *Tot Bike*, the *Skippy Racer Scooter*, the *Skippy Sno-Plane Sled*, the *Skippy Streamline Racer,* and the *Streamline Velocipede*.

Olivetti (founded 1908 near Milan) introduces the *Studio 42* typewriter designed by Alexander (Xanti) Schavinsky, a former Bauhaus student, and architects Luigi Figini, and Gino Pollini. It will set the typeform of modern typewriters for 40 years. Hermes markets an inexpensive portable typewriter, greatly reduced in size and weight, and engineered by Giuseppe Prezioso for the Swiss firm of Ernest Paillard and Company.

John Morgan heads the Sears design department, formed as a part of the Department 817, Merchandise Testing and Development Laboratory (established 1911).

Dave Chapman leaves Montgomery Ward (see 1931), and opens his own design office in Chicago. After working together at the *Century of Progress* Exposition, Jean Otis Reinecke and James Barnes also open a design office, Barnes & Reinecke, in Chicago.

Ford establishes its first styling department headed by Eugene T. "Bob" Gregorie, formerly with General Motor's Art and Color group, with a staff including Willys P. Wagner, Martin Regitko, Walter Kruke, Ed Martin, and James Lynch. Work begins on what will become Ford's first internally styled car, the 1937 Ford.

04-20. The Douglas DC-3, an airliner that can carry 21 passengers, is first flown. Designed by chief engineer James H. (Dutch) Kindelberger and designer Arthur E. Raymond (1900-1999), the DC-3 will become the first to make commercial air travel profitable and realistic. United Airlines purchases the planes from Douglas Aircraft Company and commissions interiors by Henry Dreyfuss. By 1942, 90 percent of airliners in use will be DC-3s and 11,000 will be produced throughout World War II. By the late 1980s, 2000 will still be in service.

04-21. The Standard Gas Equipment Company of Baltimore begins production of the *Oriole* stove, its first all-white model, designed by Norman Bel Geddes. The original design, conceived in 1931 as a modular system that eliminated legs and attached panels to a steel box frame, was featured in Bel Geddes's book *Horizons* (see1932) and *Fortune* magazine (see 1934). This final design approximates the original concept, but retains legs.

04-22. *Modern Living,* a line of furniture in bleached maple, designed by Russel Wright and made by Conant Ball Company for Macy's, is introduced and widely promoted. The popular line, later called *American Modern*, will make Wright famous. Wright opens a New York showroom for his work under the name Russel Wright, Inc.

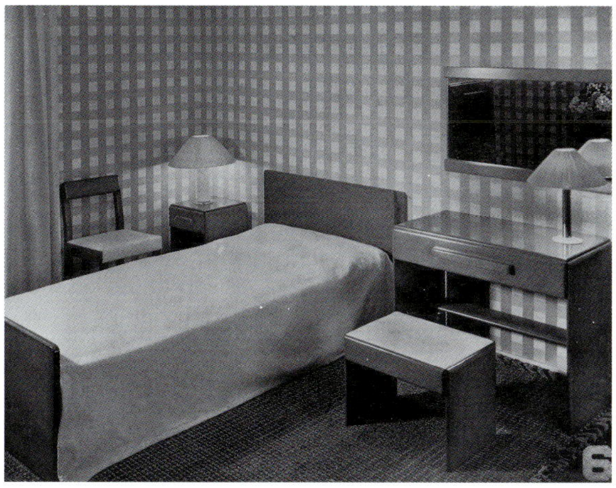

04-22

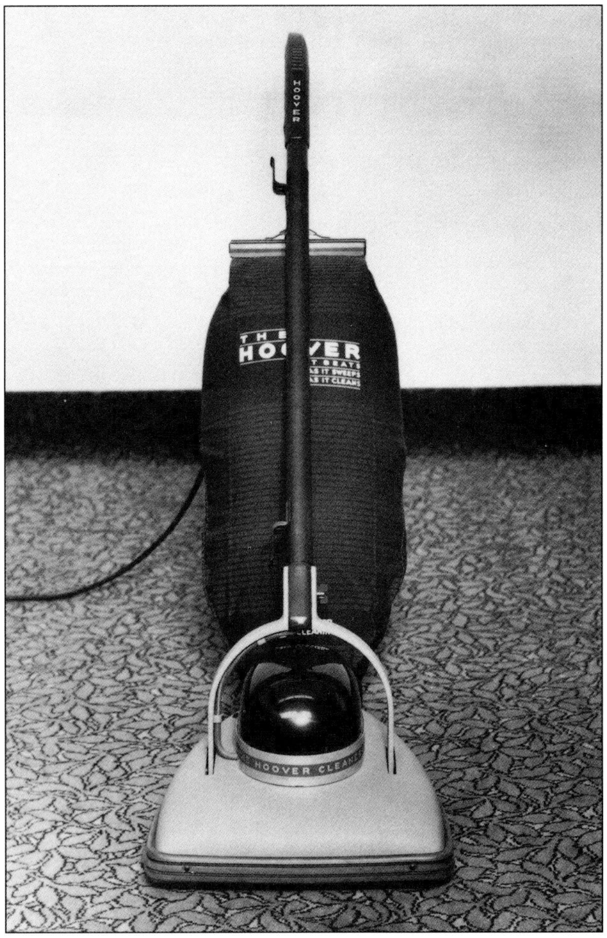

04-23

04-24

1936

Carnegie Institute of Technology (see 1934) graduates the first student in history to hold a Bachelor of Arts Degree in Industrial Design; Maud Bowers Rice, an émigré from the Philippines.

Walter Paepcke (1896-1960), president and founder of the Container Corporation of America (CCA), establishes a design department, and appoints Egbert Jacobson as design director to develop a corporate identity. CCA was founded in 1926, and developed the first cardboard packaging.

The Toledo Scale Company (see 1931) introduces the *Sentinel Duplex* scale. Toledo supported research to develop *Plaskon,* a new urea formaldehyde plastic. The scale's development began in 1934 and was completed this year by Van Doren and Rideout. It is the largest plastic molding made up to this time, and its lightweight, modernity, and compactness will increase sales by 300 percent.

Corning Glass Works introduces clear *Flameware,* a line of pressed glass cookware that includes saucepans and skillets. Glass utensils can now be used on the stovetop as well as in baking. In 1938, products with detachable glass handles attached by metal bands, such as double boilers, teakettles, and coffee percolators will be added. The line will be produced until 1979. See 1915.

Henry Dreyfuss designs a *Mercury* K-5 Pacific steam locomotive shell for New York Central Railroad's Detroit/Chicago and Detroit/Cincinnati Lines. It features cutout holes in the "white-walled" driver wheels, lit by concealed spotlights at night.

A new Harley-Davidson 61 EL motorcycle, designed by founders William Harley and the Davidson brothers, William, Walter, and Arthur, is introduced. It will later become known as the original "Hog," and will inspire the 1995 Heritage Classic model. See 1903.

04-23. The Hoover Company introduces its Model 150 upright vacuum cleaner with the first plastic hood in Bakelite (see 1909) and a magnesium body. These materials reduce the weight of the machine as well as its production cost. It is the first Hoover product designed by Henry Dreyfuss, who was engaged by the Company in 1934. His retainer fee is $25,000 per year.

04-24. Ford debuts the Lincoln *Zephyr*, designed by John Tjaarda and Howard Bonbright of the Briggs Manufacturing Company under the supervision of Henry's son, Edsel, and Bob Gregorie. Based on an earlier rear-engine design by Tjaarda shown at th*e Century of Progress Exposition* (see1933), and patented in 1935, Gregorie moved the engine to the front and added a prow. The Museum of Modern Art will later call it the first successful streamlined car in the U.S.

04-26. The Homer Laughlin Company introduces *Fiesta* dinnerware at the January Glass and Pottery Show in Pittsburgh. Frederick Hurton Rhead created the simple design in five brilliant primary colors. His consumer research discovered that users preferred to mix and match the five colors in table settings. The heavy ceramic dinnerware will be immensely popular and will become a long-lived collectible. It will continue in many colors, will still be a top seller at the turn of the century, and will develop a wide range of licensees for complementary accessories.

04-26

04-27

04-28

1937

With hopes of recreating Bauhaus ideology, László Moholy-Nagy arrives in the U.S. to become director of the School of Design at the Association of Arts and Industries in Chicago. He names the school "The New Bauhaus: American School of Design" and 20 students register. It will close in 1938 and re-open the following year as the School of Design. In 1944 it will become the Institute of Design, then become part of the Illinois Institute of Technology in 1946.

A two-year industrial design graduate program, headed by Walter Baermann, is established as the California Graduate School of Design. In 1942, it will be absorbed by the California Institute of Technology (CalTech), and renamed the California School of Design

Melamine-formaldehyde, or *Melamine*, an improved thermosetting resin developed by the American Cyanamid Company, is produced in quantity for the first time. With color stability and abrasion resistance that exceeds that of current plastics such as *Bakelite* (1909) and *Plaskon* (1936), it will be popularly used in tableware after the war.

Rohm and Haas of the U.S. introduces *Plexiglas,* a transparent polymethyl methacrylate (acrylic) sheet plastic. It will be widely used after World War II. Soon, DuPont will debut a similar material, *Lucite*.

The RCA Manufacturing Company of Camden, New Jersey, introduces a Model M portable phonograph, the *RCA Victor Special*, designed by John Vassos. The aluminum and steel phonograph features a red velvet interior.

Bendix Aviation Company introduces the first fully automatic washing machine, with a round window for viewing clothes being washed. It sells for $99.50.

Norman Bel Geddes designs a model "City of Tomorrow" of tall buildings surrounded by expressways that is featured in an advertisement for Shell Oil Company. Bel Geddes will refine this concept for the *Futurama* exhibit in the General Motors pavilion at the New York World's Fair (see 1939).

Harley Earl, head of General Motors Styling, and designer Paul A. Meyer create a concept convertible called the "Y-Job". The rounded fuselage body has suitcase fenders blended to it, and has eliminated the running board. It includes hidden headlights, push-button door handles, and electric windows. It will become the prototype for all post-war Cadillacs and Buicks.

04-29

04-27. AT&T introduces this new "300" tabletop telephone, with a receiver and transmitter in a "combined handset" that rests in a horizontal cradle. Henry Dreyfuss began work on the design in 1930. The "300" will be molded in black phenolic plastic and will be produced until 1950.

04-28. The Hobart Manufacturing Company (see 1919) introduces the *KitchenAid* Model "K" stand-mixer at the more affordable price of $55. It was designed by Egmont Arens, who had designed several earlier *KitchenAid* mixers and electric coffee grinders for the company. The appearance of the functional-looking stand mixer will remain virtually unchanged into the next century, and become a kitchen classic for serious cooks.

04-29. This Waring *Blendor,* designed by Peter Müller-Munk, is introduced at the National Restaurant Show for bartender use. The new device was invented by Frederick J. Osius and re-named after bandleader Fred Waring, who was a financial backer. In 1948, entrepreneur Hazard Reeves will buy the company from Waring, and will reformat the *Blendor* for the home market. See 1932, 1966.

04-30

04-31

04-32

1938

The American Designers Institute (ADI) is founded by a rebel group of designers who break away from the Designers Institute of the American Furniture Mart (formed 1936) in Chicago. John Vassos urges his colleagues to sever their connection with the furniture industry and expand into other products, thus forming ADI. Vassos is elected its first president. See 1944.

DuPont engineers Dr. Roy Joseph Plunkett (b. 1910) and Jack Rebok discover a synthetic non-greasy solid lubricant, the product of polymerization of tetrafluoroethylene. It will be delayed by the war, but will become known as *Teflon* (see 1960).

Adolf Hitler awards German National Culture Prizes to Ferdinand Porsche (Volkswagen), Fritz Todt (Autobahn, 1921), Ernest Heinkel (He 111 bomber, 1940), and Wilhelm Messerschmitt (Me 109 fighter plane, 1940) for their "cultural work and artistic creations."

Barnes & Reinecke of Chicago (see 1935) now has a staff including David Painter (b. 1913), vice-president in charge of design, James Teague, Fred Priess, and George Mendenhall.

The Bruynzeel Company in the Netherlands manufactures a kitchen, designed by architect Peter Zwart, featuring modular and integrated kitchen cabinets under a single counter top that will become the modern type-form.

New York Central Railroad introduces ten new streamliner steam engines and cars designed by Henry Dreyfuss for its *Twentieth Century Limited*. The new locomotives are upgraded versions of his *Mercury* K-5 Pacific locomotive shell design of 1936, and feature finned, bullet-noses reminiscent of ancient warrior helmets.

04-30. On the same day as New York Central introduces its new locomotives by Dreyfuss, Pennsylvania Railroad (PRR) introduces the world's largest steam locomotive, the S-1, designed by Raymond Loewy. The S-1 will be a feature attraction at the World's Fair (see 1939). Only one will ever be built, but it will serve as a PRR promotion, and will identify Loewy forever with the streamlined train.

04-31. McGraw Electric Company (see 1926) produces its *Toastmaster* model 1B9 (far left in photo), designed by Jean Otis Reinecke, for introduction in 1939. The first to exploit the curvature of chrome-plated shells on toasters, it will increase sales dramatically, be widely imitated, and will serve as the typeform for toasters well into the 1960s. In the photo design, progression flows from right to left, with the original Waters-Genter model (see 1926) on the far right and the 1947 IB14 model in the foreground; it will sell seven million units by 1961. In 1980, the company will change its name to *Toastmaster*.

04-32. Adolf Hitler opens the Volkswagenwerk factory in "KdF Stadt" (after War, Wolfsburg), Germany, for production of the Volkswagen Type 38 *KdF Wagen* (*Kraft durch Freude*, or Strength through Joy). Ferdinand Porsche and Erwin Komenda (1904-1966) designed it based on a prototype *Kleinauto* (see 1932). Factory construction will be halted next year as war begins. Only 210 KdF Wagens will be built for privileged Nazi officials. After the war, over 100 million of what will be called the *Beetle* will eventually be sold until production ceases in 2003. See 1932, 1948, 1949.

04-33

04-34

04-35

1939

The New York World's Fair, *The World of Tomorrow*, opens in Flushing Meadows, New York with its symbolic 700-foot tall *Trylon* and 200-foot *Perisphere* designed by Wallace Harrison, "shaped like the world with its pointer toward tomorrow." Walter Dorwin Teague is one of the eight directors of the Fair, and designer of exhibits for national Cash Register Company, Ford, Kodak, U.S. Steel, the DuPont building, the pavilion for the U.S. government, and the *City of Light Diorama* in the Consolidated Edison Building.

The Fair's Official Guidebook proclaims: "The true poets of the twentieth century are the designers, architects, and the engineers who glimpse some inner vision, create some beautiful figment of the imagination, and then translate it into valid actuality for the world to enjoy."

Henry Dreyfuss and Julian Everett designed the *Democracity* model in the *Perisphere*, which forecasts an American city in the year 2039. Dreyfuss also designed the AT&T pavilion, featuring *Vodar*, an early voice synthesizer. The General Motors building by architect Albert Kahn houses GM's *Futurama* exhibit designed by Norman Bel Geddes. The exhibit, which shows superhighways and teardrop-shaped cars of the future year 1960, is an extension of his *City of Tomorrow* model (see 1937). The theme and concept of the RCA pavilion is by John Vassos. The House of Jewels interior is by Raymond Loewy, and the Home Furnishings Pavilion is by Gilbert Rohde. In the Consumers Building, Egmont Arens designed the *Good Life* main exhibit; Gilbert Rohde the *Community Interests* exhibit; and Russel Wright the *Food Zone* exhibit. In a dramatic enactment of the future's relationship with electrical home appliances, the Westinghouse pavilion features the larger-than-life *Elektro* the robot, and his mechanical dog, *Sparky*.

04-33. *American Modern* line of ceramic dinnerware, designed by Russel Wright and manufactured by Steubenville Pottery in East Liverpool, Ohio, is introduced. It will become a great success as *the* gift for a young bride, and will sell more than 80 million pieces. Sakura, Inc., a division of Oneida, will reintroduce it in 2003.

04-34. The Radio Corporation of America (RCA) introduces its first consumer television sets, designed by John Vassos. Picture tubes, ranging in size from five inches to twelve inches, face upward at the top of the cabinet and a mirror tilts to reflect the image horizontally to viewers. This is a 12-inch TRK 12.

04-35. The first *Mercury* car debuts, conceived by Edsel Ford and designed by chief designer Bob Gregorie, with Edward A. Martin, Martin Regitko, John H. Walter, Walter Kruke, John Najjar, and Tucker Madawick.

04-36. At the World's Fair, in a speech entitled "Birth of an Industry," RCA's chairman David Sarnoff announces the first television broadcast and predicts that television will become an important entertainment and information medium. Sarnoff will credit Vladimir K. Zworykin with the invention of television. No credit is given to Philo Farnsworth (see 1920), the actual inventor to whom Sarnoff is now paying royalties, and who will die all but forgotten.

04-36

Chapter 5

1940-1949 The American Dream

As the decade opens in 1940, energized by the promises of the 1939 World's Fair, industry brings products to market to fulfill dreams of a progressive future. The American dream of rising opportunities and better living conditions is reflected in *The American Way* products offered nationally by prominent designers. The Museum of Modern Art and other cultural leaders promote new awareness of modern design that enhances a more casual interpretation of home furnishings.

However, as the war in Europe provides new employment for Americans, ends the recession, and fuels consumer purchasing power, the December 7, 1941, attack on Pearl Harbor redirects economic momentum, and U.S. manufacture of consumer products stops for four years.

Yet even in wartime, designers and manufacturers plan products and environments envisioned for the future predicted in the Thirties, and after the war the resources of the U.S. propel manufacturing capability and unimaginable growth. War production is quickly converted into peacetime manufacture of homes, cars, furniture, and domestic appliances.

Selling the new products is easy. The deferred purchases of the war years and the memories of the Depression's deprivations create an instant mass market. Almost any reasonable product is a market success. Often designs from 1941 are just reintroduced. In other cases, new technologies and materials developed for the war, such as computers, microwaves, formed plywood, plastics, and synthetic rubber are redirected into fulfilling consumer needs.

The style leader is General Motors. GM captures the futuristic dream and appeals to pride of personal ownership with exciting new models that incorporate the powerful symbolism and forms of jet planes, and rockets. Many U.S. manufacturers of home goods follow this fashion; cockpit controls, wings, and propellers begin to appear in appliances and furnishings as designers invent civilian uses for military parts and products.

Some well-known pre-war design offices have grown too costly and oversized. Their employees leave to establish their own, smaller firms. Many manufacturers establish new internal design departments. Other companies, lulled by easy markets into self-satisfaction, fail to study genuine market needs, to invest in new plants and equipment, or devise competitive design strategies.

Conditions are different elsewhere after the war, which had destroyed the manufacturing capabilities of Germany, Italy, and Japan. New factories and production processes must be built from the ground up. The largest markets are primarily in the U.S., and require U.S. distribution systems. Japan, Germany, and Italy adjust their new business strategies accordingly.

European designers do not follow the U.S. style leadership of General Motors. Their designs are more aesthetically and emotionally pleasing than Detroit's brash chrome and tail fins, and have much more in common with the simple *American Modern* furniture and dinnerware of Russel Wright's pre-war designs. New Scandinavian and Italian designs are more expressive, and seem to reflect more refined sensitivities to form, color, and design integrity than in the U.S., where design leaders recognize and publicize European products.

Design-conscious entrepreneurs in the post-war U.S. see opportunities to bring more sophisticated and affordable design to young consumers. These initiatives bring more choices to consumers, particularly young adults. The younger generation has different dreams from their parents, and responds to design benefits such as ergonomics, comfort, economy, convenience, and, yes, aesthetics.

How important is industrial design? At the end of the decade, Raymond Loewy appears on the cover of Time magazine.

05-01

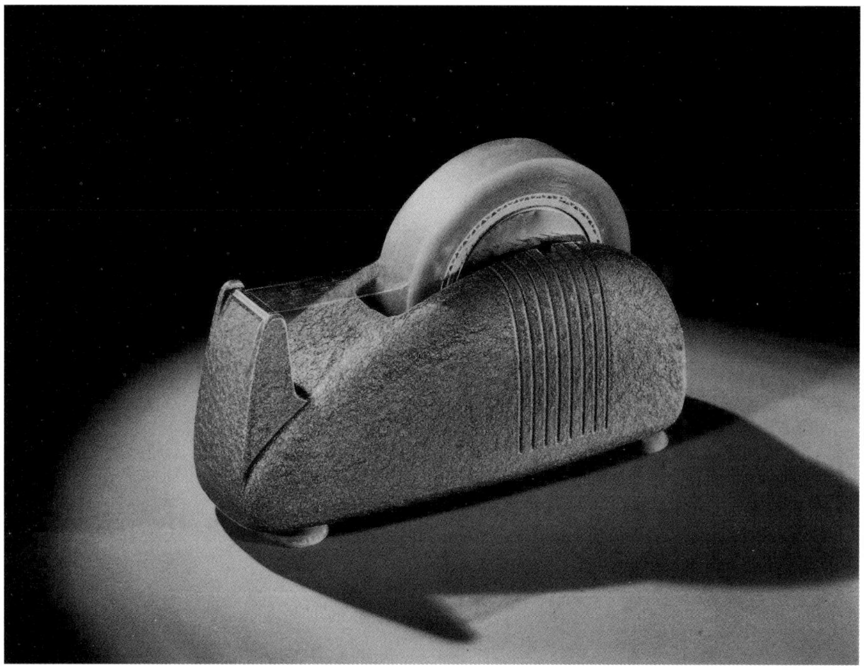

05-02

05-04

1940

Russel and Mary Wright close the home furnishings accessories business they began in 1935, and form a new company with original partner Irving Richards (1907-2003) and Eugene Morganthau, a lamp producer. Their new firm, *Raymor*, will continue to sell Wright's products, but after the war will include well-designed products by international names.

At a highly promoted patriotic event in December at Macy's featuring First Lady Eleanor Roosevelt, Russel Wright launches *The American Way,* a nationwide craft, design, and merchandising consortium of 100 designers to promote low-cost American style to industry. Due to the national preoccupation with the war, and its administrative complexity, the program will be abandoned within a year.

The Museum of Modern Art establishes a separate department of industrial design with a new director, Eliot Noyes, who was suggested by Walter Gropius. The Museum also revises the title of its traveling exhibition to *Useful Objects of Design Over $10.*

The German wartime counterpart to the U.S. *Jeep* (see 1941) is the Volkswagen *Kübelwagen* (Bucket car). It will be revived in 1984 as Volkswagen's *Thing.* An amphibious version, the *Schwimmwagen* or *Schwimmkübel,* is also built. Ferdinand Porsche (see 1932) designed both.

The Curtiss-Wright P-40 *Tomahawk* fighter, the first U.S. plane to be produced in very large quantities (14,000 will be built), enters service for the Royal Air Force (RAF). It was first flown in 1938 as the XP-40, developed by a team led by project engineer William O. Watson. Later versions will be known as the *Warhawk* (for the U.S. Air Force-USAAF) or *Kittyhawk* (RAF).

Consolidated Aircraft's B-24 *Liberator* four-engine bombers, designed under the guidance of Chief Engineer I. M. Laddon, are produced within nine months, and used to fly war supplies to England. Within a year at a new Willow Run aircraft plant, Ford Motor Company's Charles ("Cast Iron Charlie") Sorenson will increase Ford's production rate through the first application of automotive "sub assembly" techniques to aircraft. The bomber will become one of the most destructive weapons of the war, and 18,500 will be produced.

05-01. Lincoln's new Zephyr *Continental* Cabriolet is designed by Eugene "Bob" Gregorie with Edsel Ford's financial support. The design will continue through the 1948 model year, becoming a classic and winning acclaim of the Museum of Modern Art (see 1951).

05-02. 3M (see 1931) introduces a heavy cast metal desktop tape dispenser in gray (shown) by Jean O. Reinecke of Barnes & Reinecke. It replaces his first 1938 design for 3M, and he will redesign it again in 1953. Reinecke also redesigns a disposable plastic Scotch tape dispenser that will become one of the most successful and enduring examples of 20th century design. It is an ingenious refinement of one he designed last year, of stamped sheet metal. All will still be in production in 1998. See 1961.

05-04. Brooks Stevens redesigns this Edmilton Corporation's *Steam-0-Matic* iron, first introduced in 1936. His streamlined design decreases weight and boosts sales by 15-fold in a month.

05-05

05-06

05-08

05-09

1941

The Museum of Modern Art holds a competition conceived by Edgar Kaufmann, Jr., and organized by Eliot Noyes, to discover imaginative designers of contemporary furnishings. Prizes are awarded to Charles Eames and Eero Saarinen for chairs and storage pieces. Competition entries are exhibited by the Museum as *Organic Design in Home Furnishings.*

80 percent of U.S. households are wired for electricity. 79 percent own electric irons, 52 percent have electric washing machines and refrigerators, and 42 percent have vacuum cleaners.

Blenko Glass (see 1921) engages its first professional designer, Winslow Anderson. William H. Blenko, the son of Blenko's founder, headed the company after his father's death, and had designed most of Blenko's blown pieces to date.

The U.S. Army adopts the deep rounded contour M-1 steel helmet for military use. The helmet was developed by Bashford Dean, a U.S. ichthyologist and paleontologist who had worked on the technological history of armor as a Major in the U.S. Army after World War I, and later became curator of arms and armor at the Metropolitan Museum of Art.

05-05. Corning Glass Works produces, in Pyrex® (see 1915), this *Chemex* coffeemaker invented by Dr. Peter Schlumbohm (1896-1962). He derived the design from an ordinary Erlenmeyer chemical flask, funnel, and paper filter. It will become an instant success and inspire the design of future coffeemakers.

05-06. The American Bantam Car Company (see 1929) of Butler, Pennsylvania, with consulting engineer Karl K. Probst, develops the first successful design for a General Purpose (GP) army vehicle. Willy's engineer Delmar G. "Barney" Roos perfected this design based on the 1940 Willys-Overland *Americar,* the lightest full-sized car on the road. Standardized by the Army as the MB, contracts will be awarded to Bantam (Bantam BRC, *Blitz Buggy),* Willys (Willys MA, *Quad*) and Ford's GPW (Government P-class Willys, named the *Pygmy).* The "GP" designation leads to its popular name, the "Jeep." More than 600,000 will be built during the war. See 1945.

05-08. Chrysler exhibits the concept car, *Thunderbolt*, the first streamlined car with integrated fenders. Ralph S. Roberts, manager of Briggs-LeBaron, Chrysler's major body producer, patents the design, but Briggs-LeBaron designer Alex Tremulis was the creative source.

05-09. Packard Motor Car Company introduces its new *Clipper* car by Werner Gubitz, head of Packard design, and Howard "Dutch" Darrin, with contributions from Alex Tremulis and Edward Macauley, Packard Vice President. The *Clipper's* design is very competitive with the big three, but hopes for commercial success will be dashed by the start of the war. It will be re-introduced in 1946 virtually unchanged (1946 model illustrated).

05-10. Edmilton Corporation introduces this dual-soleplate *Petipoint* clothes iron designed by Brooks Stevens. Stevens chose the name because of the "petite point" at the rear for ironing small details. The streamlined design features horizontal cooling fins and an upswept tail.

05-10

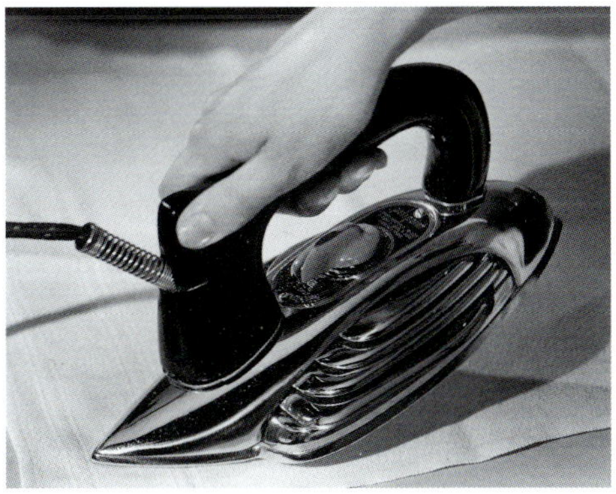

05-11

05-12

05-13

1942

The Container Corporation of America (CCA) introduces its Color Harmony Manual, a color specification system of 944 movable color chips based on the classification system devised by Wilhelm Ostwald in 1916. Developed by Egbert Jacobson, CCA's design head, with physicists Walter C. Granville and Carl Foss, the system will be adopted widely as a product designer's tool for consistent color specification.

The standard weapon used by the U.S. Army is the M-1 Springfield .30 caliber semi-automatic rifle. Canadian-born John Cantius Garand (1888-1974) of the Springfield Armory designed the new rifle that was adopted in 1936 to replace the previous bolt-action Springfield, now more than 40 years old. Four million will be produced in the next four years.

George Nissen signs his first client, The Army Air Corps, for the device he invented and developed during the last few years. He names his invention a "trampoline," after "el trampolino," "diving board" in Spanish. Nissen, a college diver, wanted to simulate the free fall sense of diving. The Air Corps uses the trampoline to train airmen in the sensation of weightlessness during flight maneuvers. After the war, the trampoline will become a popular sports exercise device, and, by 2000, will become an Olympic event.

U.S. Rubber Company achieves structural integrity of fiberglass by using a new polyester resin developed by the Columbia Chemical Division of Pittsburgh Plate Glass Company and fibrous glass fabric from Owens-Corning Fiberglas. The new material is used in bullet shields for aircraft gas tanks and military helmet liners. Polyethylene (see 1933,1947) is first produced commercially, and the first plastic bottles are blow-molded. DuPont produces *Lucite®* (see 1937) for use in bomber nose cones. Thermoformed plastic blister packaging is developed.

05-11. Lockheed Aircraft Company's new P-38 *Lightning* fighter plane is designed by Clarence "Kelly" Johnson. It has three fuselages: the middle with seating for the pilot, and right and left fuselages with engines, fuel, and individual tail fins. The design will inspire Harley Earl's post-war automotive tail fins. Johnson will organize the "Skunk Works" next year, Lockheed's top-secret high-tech development group, to reduce bureaucracy and produce rapid results. Johnson is credited with the famous acronym, K.I.S.S. ("Keep it Simple, Stupid").

05-12. The first prototype rocket-propelled guided missile, the AS-4, completes a successful flight. Herman Oberth, working under Wernher von Braun, began developing the missile in 1934, based on Goddard's patents (see 1926). Oberth perfected the controlled burning of liquid hydrogen and oxygen stored in vacuum flasks. The missile weighs 12 tons, is 46 feet long, and reaches a speed of 3,300 mph. The rockets will be built at Nordhausen, Germany, will be named *Vengence Weapon 2* (V-2), and will bomb London in 1944.

05-13. The Boeing four-engine B-17C bomber, developed in 1936 and supplied to the RAF in 1940, is now called to duty for the U.S. Named the *Flying Fortress*, it will be a mainstay for the Air Force during the war in both Pacific and European theatres, and more than 12,000 will be produced.

05-14

05-15

05-16

1943

Donald Dohner (see 1935) and J. Gordon Lippincott establish the New York product and package design firm of Dohner & Lippincott in partnership with the Douglas T. Sterling Company of Stamford, Connecticut, a manufacturer who provided financial support.

Interiors magazine (first published in 1940) initiates a regular section called "Industrial Design," conceived and edited by Donald Dohner, which features manufactured products. Budd Steinhilber illustrates the cover of the first issue carrying the new section, and both Steinhilber and Read Viemeister illustrate many of Dohner's design concepts that are the subjects of his articles. Upon Dohner's sudden death in December, his business partner J. Gordon Lippincott becomes editor. The section will expand and become *Industrial Design* magazine (see 1954).

Peter Müller-Munk is commissioned by Dow Chemical Company to stimulate interest in plastics for post-war products. He uses the opportunity to explore the idea of a kitchen as a domestic production line that combines factory efficiency with "charm and livability." Müller-Munk also recommends, and Dow promotes, *Durez* plastic housings for lighter weight sewing machines.

The technical design discipline of Human Factors begins as the armed services conduct research in order to reduce accidents caused by equipment users' human error. The measurement of human dimensions results in standards for the design of military equipment. Such data will form the basis of post-war design standards for consumer products published by Henry Dreyfuss as the "Joe and Josephine" body measurements in 1955 (see 1960).

Joseph Gerber, director of an Oregon advertising agency, works on the manufacture of steel knives with an integrally cast aluminum handle, a process which was developed in the 1930s by mechanic David Murphy. Gerber and Joe Pollock, his agency's art director, redesign the elegantly shaped handle, which will be selected by the Museum of Modern Art after the war for its "Good Design" exhibitions (see 1950), and the knife will become known as "Gerber Legendary Blades."

05-14. Galvin Manufacturing Corporation, later named Motorola, Inc., developed this *Handie-Talkie*™ in 1940 as a lightweight, handheld, two-way radio that is now being widely used during World War II. Recognized around the world, the *Handie-Talkie* radio will often be perceived in the future as the grandfather of portable communications devices. See 1983.

05-15. Before entering the Army, Jens Risom designs this chair of simple wood construction and surplus military webbing for Hans Knoll. This chair will be the only modern furniture available during the war, and will be patented in 1945.

05-16. The Lockheed C-69 military transport plane is the first to fly non-stop coast-to-coast. Conceived by Howard Hughes and designed by Charles "Kelly" Johnson in 1939 as a commercial plane, it will be reconverted for use by Trans World Airways (TWA) as the *Constellation*, entering commercial service in 1947 with interiors designed by Alfons Bach. With its triple tail and graceful lines it will be considered the most beautiful airliner of the day and will become the last plane ever flown by Orville Wright (1871-1948).

05-17

05-18

05-19

1944

In February, the Society of Industrial Designers (SID) is established by 15 practitioners: Walter Dorwin Teague, president, Henry Dreyfuss, vice president, Raymond Loewy, chairman of the executive committee, Harold Van Doren, treasurer, Egmont Arens, secretary, and members Donald Deskey, Norman Bel Geddes, Lurelle Guild, Ray Patten, Joseph Platt, John Gordon Rideout, George Sakier, Joseph Sinel, Brooks Stevens, and Russel Wright. Each of the founding members will invite one additional designer to join next year. Membership requires evidence of the design of at least three mass-produced products in different industries. The Society is formed in part to reinforce the legality of industrial design as a profession. New York State court had found Teague innocent in a dispute over his payment of 1941 business taxes. This decision implied that industrial designers were professionals rather than merchants.

The American Designers Institute (ADI—see 1938, 1951), led by Alfons Bach, Ben Nash, and George Kosmak, relocates its administrative office to New York from Chicago. One of the group's first projects is the development of a prototype industrial design education program based on a survey of its members; this will become the model for most post-war design programs.

The Council of Industrial Design (CoID) is established in London, to foster good design for the products of British industry. See 1956.

Richard F. Bach, of the Metropolitan Museum of Art, hosts a conference of design school administrators to examine industrial design education. A summary states, "the definition of the term Industrial Design was carefully avoided, fearing such action might constrain discussion." Subsequent meetings will lead to a formal organization of educators (see 1948).

05-20

05-17. The Boeing B-29 *Superfortress* enters action in the Pacific theatre. It is the first bomber with pressurized crew compartments, and with a central machine-gun fire control system. The *Superfortress* will become part of history when it drops the atomic bombs on Japan in 1945.

05-18. Donald Deskey designs this original concept of a bowling center ball return for Brunswick-Balke-Collender Company for post-war recreation.

05-19. The German Messerschmitt *Me 262* fighter jet enters service in the Luftwaffe with a jet engine developed by Hans J. Pabst von Ohain in 1939. One of the fastest fighter aircraft of the war at 560 mph (901 km/h), it is also the first operational jet fighter, has the first swept-back wings, and is the first to use air-to-air missiles.

05-20. In response to a U.S. Navy request for "seaworthy" furniture, Witton C. Dinges, the founder/president of the Electric Machine and Equipment Company (Emeco) of Hanover, Pennsylvania, develops a line of sturdy aluminum furniture that includes this seven-pound square-tubed chair with contoured seat and slat-back known as model 1006. It will become standard post-war equipment in government offices, prisons, and hospitals and will be revived in 1990 for chic restaurants. By 2000, half a million will have been sold, and Emeco will introduce "Heritage" models based on the Model 1006, as well as more exotic versions designed by Philippe Starck to be sold at The Terence Conran Shop in New York (see 1977).

05-21

05-22

05-23

1945

After the death of Donald Dohner last year, J. Gordon Lippincott, and Walter Margulies begin their successful association, which will result in the firm of Lippincott & Margulies. The firm will still be a prominent design presence well into the next century.

Harley J. Earl establishes an independent styling studio, Harley Earl, Inc., to design non-automotive products. In the mid-1950s its name will be changed to Harley Earl Associates, Inc., and in 1964 will become Ford & Earl Design Associates.

George Nelson and Henry Wright, co-editors of *Architectural Forum*, design a *Storage Wall* concept as an outgrowth of their joint book just published, *Tomorrow's House*. The *Storage Wall* is featured in *Life* magazine and attracts much public attention. It provides a contemporary alternative concept to closets, and leads to Nelson's association with the Herman Miller Furniture Company. In 1949, the Company will produce Nelson's *Basic Cabinet Series* in a realization of the concept.

Don Dailey of Harold Van Doren's office designed the classic Boston pencil sharpener just introduced. It will be used in millions of schoolrooms for the next 50 years.

Philadelphia shipyard engineer Richard James observes a torsion spring fall off his desk, and develops the *Slinky,* a spring-like toy. It is first sold at Gimbel's in Philadelphia for $1.00. By the end of the century, 250 million *Slinkys* will have been sold, and a *Slinky* stamp will be issued.

The first commercial ballpoint pen, designed by J. Gordon Lippincott & Company for Milton Reynolds, a Chicago salesman who had seen the Biro pen while vacationing in Argentina, is introduced for $12.50 at Gimbels'. It sells out the entire stock of 10,000 on the first day. Lower costs and mass distribution will not be achieved until French licensee Marcel Bich introduces his *Bic* pen in 1953.

A CJ-2A *Universal Jeep* designed by Barney Roos goes into civilian production at the Willys-Overland plant in Toledo, Ohio, even as military production continues. It sells for $1,090. *Jeeps* will be produced by Willys-Overland until 1953, when the Company will be bought by Henry J. Kaiser and be renamed Willys Motors. Renamed the Kaiser Jeep Corporation in 1963, it will be acquired by American Motors Corporation (AMC) in 1970. AMC will cease production of the CJ in 1985, only to be forced by public outcry to resume. In 1986, Chrysler will buy the name and line along with the rest of American Motors, and the *Jeep* will go on. Military *Jeeps* will continue to be produced until 1982. See 1941.

05-21. Castleton China Company debuts *Museum Service* white dinnerware designed by Eva Zeisel for the Museum of Modern Art, as "the first Modernist porcelain dinnerware in the U.S."

05-22. Former E.I. DuPont chemist Earl Silas Tupper (1907-1983), who has designed and is producing these seven-ounce polyethylene bathroom tumblers, founds Tupperware Corporation. He started business in 1938 as a custom molder in Farnumsville, Massachusetts, and developed his manufacturing technique in 1942. See 1947,1994.

05-23. Hamilton Beach Manufacturing Company introduces this Model E stand mixer. After the war pent-up demand for manufactured goods unleashes a torrent of new consumer products. Demand exceeds capacity for months.

05-24

05-25

1946

Buckminster Fuller builds a prototype of his circular *Dymaxion House* in Wichita, Kansas, with components produced by Beech Aircraft Company. Conceived as a solution for global affordable housing, this version included his prefabricated *Dymaxion Bathroom* concept from 1937. It was patented in 1942 in response to the aircraft industry's demand for low-cost temporary housing. Despite orders for 35,000 units, Fuller will delay production with design improvements; potential buyers find other options and his company will go bankrupt.

Edgar Kaufmann Jr. is appointed director of the Museum of Modern Art's department of industrial design, replacing Eliot Noyes (see 1940). The Museum of Modern Art presents the work of Charles Eames in the exhibition, *New Furniture Designed by Charles Eames.*

The "Do-it-Yourself" phenomenon begins with the first low-cost consumer power tools. Black & Decker introduces its first consumer power tool, a red *Home-Utility* 1/4" portable electric drill, priced at $16.95. Albert Kaufman, inspired by the sewing machine, invents the hand held "jig-saw".

Walter Gropius and several of his colleagues at Harvard University, including architect Benjamin Thompson (1918-2003), found The Architects Collaborative (TAC) to promulgate modern design in the postwar era. Thompson will establish a retail shop, *Design Research, Inc.* (see 1951).

At the Housewares Show in Atlantic City, only two exhibitors are in the plastics business, Tupper Plastics Inc. (see 1945) and Beacon Plastics Corporation. The American Hardware Manufacturers Association (AHMA) initiates an annual National Hardware Show.

Hans Knoll and Florence Schust found Knoll Associates in New York in order to create a "collection" of furniture pieces by well-known designers. In 1959 Knoll will be purchased by Art Metal, which in 1969 will merge with Heller Inc. and become Knoll International. In 1977, General Felt Industries will acquire Knoll, and later it will be bought by Westinghouse (see 1990).

05-24. Wurlitzer introduces this Model 1015 jukebox designed by industrial designer Paul Fuller. With chrome trim, garish colors, and bubbling lights, it will become a classic and ultimate cultural typeform. Wurlitzer will produce jukeboxes until the product loses popularity in the 1960s, and will end U.S. production in 1974, but its German subsidiary will continue production into the late 1990s. See 1900.

05-25. Ekco Housewares Company, founded in 1888 by Edward Katzinger, introduces *Flint 1900,* a line of kitchen tools with *Bakelite* handles designed by Jack Zimmer, Walter Luneberg, and James Hvale of the Ekco staff.

05-27. General Mills Inc. introduces its first consumer product, the Betty Crocker *Tru-Heat* iron, Model GM1A, that can rest on either side, rather than its heel, and whose double-pointed shaped soleplate can iron forward or backward. Designed by John N. Polivka (1910-1999) of General Mills's Appliance Division, and consultant designer Francesco Collura, the iron will catapult General Mills to a top competitive position. By 1954, the company will exit the consumer business and its appliance division will be sold to McGraw Electric Company.

05-27

05-28

05-29

05-30

1947

At Bell Labs, William Shockley (1910-1989), John Bardeen (1908-1991), and Walter Brattain (1902-1987) invent the transistor, an efficient and inexpensive low voltage switch. By 1997, more than half a billion transistors will be manufactured every second.

The Society of Industrial Designers (see 1944) organizes an exhibition of industrial design photographs that opens in Philadelphia, and travels to 12 other cities. The display presents the work of 32 designers; it is the first time U.S. practitioners exhibit their work in friendly competition.

Business consultant W. Edwards Deming (1900-1993) advises Japanese industrialists to utilize innovative methods, and focus on the quality of workmanship and functional reliability. The Japanese heed his advice, but the U.S. will not until the 1980s, when his theories will become known as Total Quality Management (TQM). By then Japan will have captured many U.S. markets.

Gordan Florian, an industrial designer in Fairfield, Connecticut, begins teaching an evening program of industrial design at the University of Bridgeport, primarily for General Electric personnel. In 1949, the University will initiate a four-year degree program in industrial design, headed by Robert Redmann.

The National Housewares Manufacturers Association (NHMA) holds the first large "Housewares Show" in Philadelphia Convention Hall. Earlier shows had been held at hotels in Chicago since 1927, and in New York since 1932.

Polyethylene refrigerator food storage containers made of "Poly-T" by Earl C. Tupper are featured in *House Beautiful* under the title, *Fine Art for 39 cents.* The containers, known as *Tupperware,* will later be placed in the permanent collection of the Museum of Modern Art. By 1951, "Tupperware Home Parties," pioneered by Tupperware's marketing dynamo Brownie Wise, will distribute the product nationwide, and become a multi-million dollar business. See 1945, 1994.

05-28. Coca-Cola debuts this new soda-fountain dispenser designed for Dole Deluxe by Raymond Loewy of Raymond Loewy and Associates.

05-29. Studebaker Corporation introduces the first genuine newly designed post-war car that Eliot Noyes praises as "so refreshing it made other cars look out of date." Raymond Loewy Associates began the design in 1939 with Virgil Exner heading Loewy's team of Robert E. Bourke and modeler Frank Alhroth. In 1944, Exner and Studebaker vice president Roy E. Cole conspired to bypass Loewy, who learned of the scheme and fired Exner, naming Bourke to replace him. Exner was promptly hired by Cole to finish the work.

05-30. Herman Miller Furniture Company produces these molded plywood chairs with rubber shock mounts designed by Charles and Ray Eames. The original concept, designed by Eames in collaboration with Eero Saarinen, won an award from the Museum of Modern Art at an exhibit before the war (see 1941). Early LCW (Low Chair-Wood) and DCW (Dining Chair-Wood) designs were first produced in 1946 by Evans Products Company (Eames' wartime employer) and are now produced and distributed by Herman Miller. Produced until 1957, the chairs will be re-issued in 1994. The *Eames Chair* will be selected as "best design of the century" by Time magazine in 1999.

05-31

05-32

1948

The Museum of Modern Art holds an international competition for *Low Cost Furniture Design* for furniture retailers agreeing to produce winning designs. One chair by Charles Eames, with an organically shaped one-piece stamped metal bucket seat, will be produced in fiberglass (see 1951). The Museum of Modern Art consolidates its department of industrial design under Edgar Kaufmann, Jr., and its department of architecture under Philip L. Goodwin, both now reporting to Philip Johnson.

House & Garden magazine begins its Color Program to identify annual market color trends, and provide forecast specifications for manufacturers to meet the new post-war consumer demand for color in products. Faber Birren, an internationally recognized color authority, is named as the program's consultant. The program will continue until the 1980s.

Richard F. Bach, Dean of Education of the Metropolitan Museum of Art (see 1944) initiates meetings to establish the National Association of Schools of Design, with twenty-two member schools. The organization accredits art and design education programs, and will be renamed the National Association of Schools of Art and Design (NASAD) in 1981.

Acrylonitrile-butadiene-styrene (ABS) structural plastic is developed by Otto Bayer in Germany, and initially used in luggage.

The Volkswagenwerks in Germany, largely destroyed by 1944 bombing, is offered for sale to, among others, the Ford Motor Company. Henry Ford II declines the offer, as "not worth a damn". Conditions of the Cold War return ownership to Germany, and Heinz Nordhoff, pre-war executive for Opel, becomes director of the Wolfsburg plant. Volkswagen will become West Germany's major postwar industry. See 1932, 1938, and 1949.

05-33

05-31. Knoll Associates (see 1946) introduces the *Womb Chair,* designed two years earlier by Eero Saarinen as "a great big chair you can curl up in." The chair's design stems from an earlier design of his, created in collaboration with Charles Eames (see 1941).

05-32. Preston Thomas Tucker (1903-1957), introduces the *Tucker 48* "Car of Tomorrow." The original 1946 *Torpedo* design by George Lawson had three headlights, the center one of which turned along with the turn of the steering wheel. Tucker engaged Alex Tremulis and a design team from Lippincott & Margulies (L & M) (see 1945) that included Hal Bergstrom, Philip S. Egan, Tucker Madawick, Budd Steinhilber, and Read Viemeister. The prototype, named the *Tin Goose*, used Tremulis's body design and the front and rear designed by the L & M team. In 1949, the SEC will indict Tucker on 31 counts of fraud, theft, and regulatory violations, and close production after a run of 51 cars. Tucker will be acquitted in 1950. In 1988, the feature film, *Tucker: A Man and His Dream*, directed by Francis Ford Coppola, will be released.

05-33. This *Polaroid* camera Model 95, which develops its own prints in minutes, is first sold. The camera was invented by Edwin H. Land (1909-1991) who got the idea while on vacation in 1943, and his daughter asked to see the picture he had taken. Priced at $89.75, the camera was designed by Walter Dorwin Teague Associates.

05-34

05-35

05-36

1949

Popular Mechanics forecasts that, "Computers in the future may weigh no more than 1.5 tons."

Raymond Loewy is featured on the cover of *Time* magazine, the first time an industrial designer has achieved such recognition. Loewy has a staff of 200 working under partners William Snaith, A. Baker Barnhardt, and John Breen. Employees who work without much recognition include Richard Latham, Peter Thompson, Joseph Parriott, Jay Doblin, Gordon Buehrig, and Tucker Madawick. Loewy, with his yearly gross income now exceeding three million dollars, compared to $1 billion in yearly sales of Loewy-designed products, concentrates on building his own name and image.

The *Hasselblad 500C* reflex camera, designed by Swedish designer Sixten Sason (1912-1969) in 1947 for Victor Hasselblad, is introduced. In 1985, the design will be voted the "best international design classic" of 1900-1960 by registrants of the *WORLDESIGN 85* Conference.

A unique "ball" wall clock with individual projecting balls radially arranged to represent each hour, is manufactured by the Howard Miller Clock Company, and is sometimes called the "atomic" clock. It is the first of many modern clock designs by George Nelson for the company, founded in 1927 by D.J. DePree as the Herman Miller Clock Company. Its name changed in 1937 when Herman's son Howard took it over. The company also manufactures plastic "bubble lamps" designed by Nelson.

Bel Pon of the Netherlands, the only international distributor of Volkswagenwerks, introduces a Volkswagen car produced in Germany (see 1948) in the U.S. Few will notice, but U.S. sales will grow to 330 in 1955, and 79,000 by 1957. The car will soon be called the *Beetle*. See 1932, 1938.

05-34. Henry Ford II chooses this 1949 Ford design by independent designer George W. Walker over a design by Ford Styling Chief Eugene "Bob" Gregorie. The design will be very successful, will sell more than a million, and will lead to Walker replacing Gregorie (see 1955). The design originated from a clay scale model submitted to Walker by his new employee, Dick Caleal, whose model was selected from three competing models, one by Joe Oros and Elwood Engel of Walker's office, and the third by Gregorie. Caleal, with Walker's permission to work at home, built the clay model on his kitchen table. The aircraft inspired "spinner"-type grill had appeared in Ford concept models since 1946. In 2003, the Ford Motor Company will officially recognize Dick Caleal as the designer of the 1949 Ford.

05-35. This Model 500 desk telephone, designed by Henry Dreyfuss Associates, is put into service by AT&T. In 1954, the 500 will be the first model to be offered in colors other than black and will still be the most commonly used phone in the U.S. in 1995.

05-36. The Herman Miller Furniture Company introduces these *Basic Storage Components,* designed by George Nelson and based on his earlier concepts (see 1945). The furniture provides necessary home storage space and can also be used as a space divider. Herman Miller also introduces *Storage Units* designed by Charles and Ray Eames.

GH4
LKJ 5
MNO 6
PRS 7
0
OPERATOR

Chapter 6

1950-1959 Good Design

The Fifties herald a fresh atmosphere of change in design activity. The Museum of Modern Art crusades for "Good Design" with promotions for the general public and business leaders. The Industrial Designers Institute initiates annual design awards, and the Society of Industrial Designers exhibits U.S. designs nationally and internationally. Italy establishes the *Compasso d'Oro* (Golden Compass) design awards. Conferences seek to identify and articulate clear objectives for the design profession. Design magazines appear for the first time in Italy, England, and the U.S. The journals circulate internationally, increasing the awareness of a global design community and international standards of design excellence.

Some U.S. manufacturers, influenced by European companies such as Olivetti, become aware of design as a competitive and strategic business tool, and dismiss the superficial theme-styling of the Thirties grand expositions and post-war General Motors cars. In furniture, Herman Miller and Knoll open windows to new thinking about interior design.

The outlines of a growing global economy emerge at the end of the post-war period. Germany, Japan, and Italy have rebuilt their industrial capabilities and their automobiles, furniture, and electronic products compete in U.S. markets. Design standards, communicated by new international design organizations and magazines, and fostered by the Museum of Modern Art, have solidified into what many term an "international Style" typified by functional, undecorated organic or geometric forms. In fact, this style is a reincarnation of Bauhaus design principles, led by the "New Bauhaus" school at Ulm and Braun AG of Germany.

U.S. automotive design, on the other hand, led by the immense scale and post-war futuristic vision of General Motors, intensifies its glittering chrome trim and jet-plane symbolism through the decade. This style finds its way into kitchens, major appliances, and small appliances, as if they were automotive after markets. Companies continue to promote "Kitchens of the Future," "Houses of the Future," and "Dream Cars," but this trend subsides after the technological shock of *Sputnik* in 1957 introduces a new understanding of the present.

The very public failure of the *Edsel* car and the dramatic loss of market share to many imports, further reveal U.S. insensitivity to international design directions. Many successful import designs such as the Volkswagen *Beetle*, Braun appliances and Sony's transistor radios, however, offer popular design choices to consumers.

Competition changes the balance of the market. Major auto names such as Hudson, Nash, Packard, and Studebaker disappear, and Detroit begins to consider smaller, less ostentatious car designs. High-volume products like small appliances bring decreasing profit margins, and large-scale manufacturers force out many smaller competitors. Late-decade U.S. international successes, however, include Fuller's geodesic dome and Boeing's 707 jetliner.

By the end of the decade, the pre-war Depression, World's Fair futurism, nationalism, and post-war material satisfactions no longer dominate U.S. market attitudes. Young adults are more sophisticated, more internationally minded, and demand higher product quality. Global competitors understand this trend and design accordingly. However, many U.S. manufacturers still do not, and suffer the economic consequences.

06-01

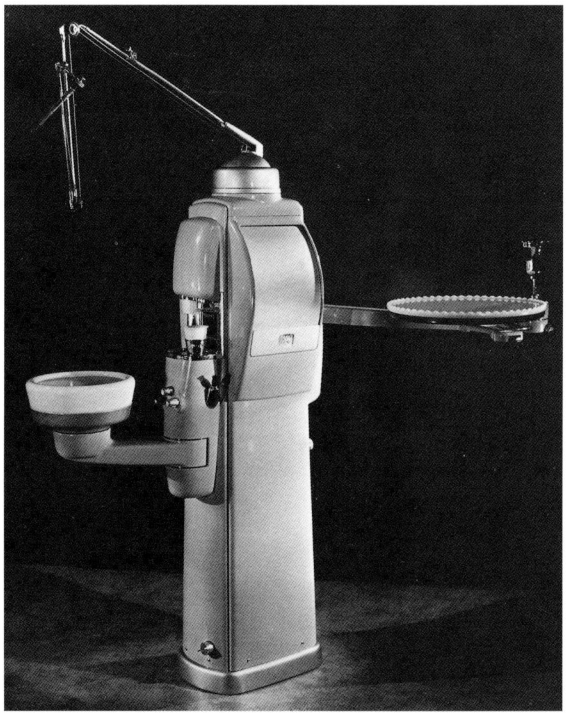

06-02

06-03

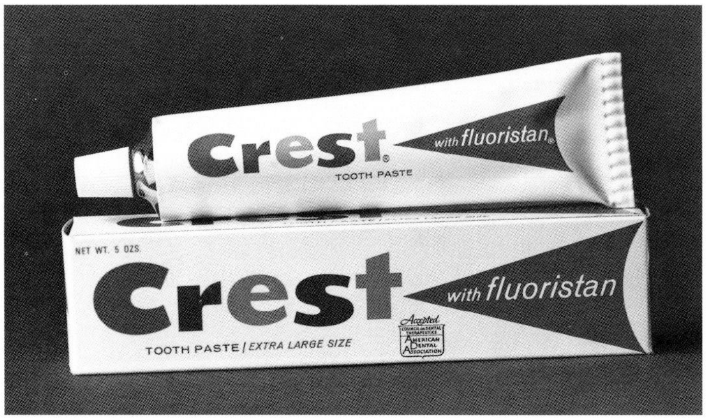

1950

Formica (see 1913) becomes the material of choice in post-war home decor. The *Skylark* pattern, with thin-line boomerang shapes, is designed by Brooks Stevens Associates. Nettie Hart of Raymond Loewy Associates will refine the pattern in 1957. An admired 1950s shape, boomerangs suggest supersonic airplane wings and modernity.

Edgar Kaufmann Jr. of the Museum of Modern Art launches a *Good Design* exhibition in collaboration with the Merchandise Mart in Chicago. René d'Harnoncourt, director of the Museum, and Wallace O. Ollman, general manager of the Mart, issue a statement about the exhibition that says in part "Two national institutions believe and hope that in combining their resources they will stimulate the appreciation of and creation of the best design among manufacturers, designers, and retailers for good living in the American home". Nearly 250 products are selected on the basis of "eye appeal, function, construction, and price, with emphasis on the first," and are promoted with "Good Design " hangtags designed by Morton Goldsholl. The exhibitions will continue annually until 1955.

Libbey Glass, Inc., debuts a line of *Classic Crystal* beverage glassware designed by Freda Diamond that is recognized by the Museum of Modern Art in its annual "Good Design" exhibition.

The New York and Philadelphia design offices of Van Doren, Nowland & Schladermundt announce that they will operate as two independent firms. Van Doren remains in control of his Philadelphia office, Harold Van Doren and Associates, and the New York firm will be known as Nowland and Schladermundt until 1955.

Planner Group furniture, designed by Paul McCobb and made by the Winchendon, Massachusetts, Furniture Company, is introduced as the first affordable system of modular furniture. McCobb will be extremely influential in popularizing a new American style of interior design characterized by furniture with low, clean lines and trim foam cushions, and in the use of modular furniture and room divider wall systems.

Sunbeam Corporation introduces the T-20 Radiant Control Toaster, an automatic two-slice chrome toaster with innovative transverse slots patented by consultant designer Robert Davol Budlong, and designed in cooperation with staff designer Ivar Jepson. The design will remain on the market until 1995.

Rosti of Denmark introduces classic *Margrethe* melamine nesting mixing bowls designed by Bernadotte & Bjørn. The design will be produced in the U.S. by Copco Incorporated (see 1960), and set the typeform for plastic bowls for the rest of the century.

06-01. Walter Dorwin Teague Associates completes work on this *Instru-Matic Dental Unit* for the Ritter Company, Inc. It will be introduced next year.

06-02. The Porsche 356 is produced. Designed by Erwin Komenda (see 1938) with elements of the Volkswagen, it was prototyped in a Gmünd, Austria, factory in 1948. Upon the death of Dr. Ferdinand Porsche (see 1932) next year, the Porsche firm will be headed by his son, Ferdinand "Ferry" Porsche II (1909-1998). He will introduce the 356 in 1952, which will quickly become a classic, with successive designs such as the Porsche 911 in 1964, the *Targa,* and the mid-engine 914 in 1974.

06-03. Donald Deskey designs this *Crest* toothpaste package for Proctor & Gamble.

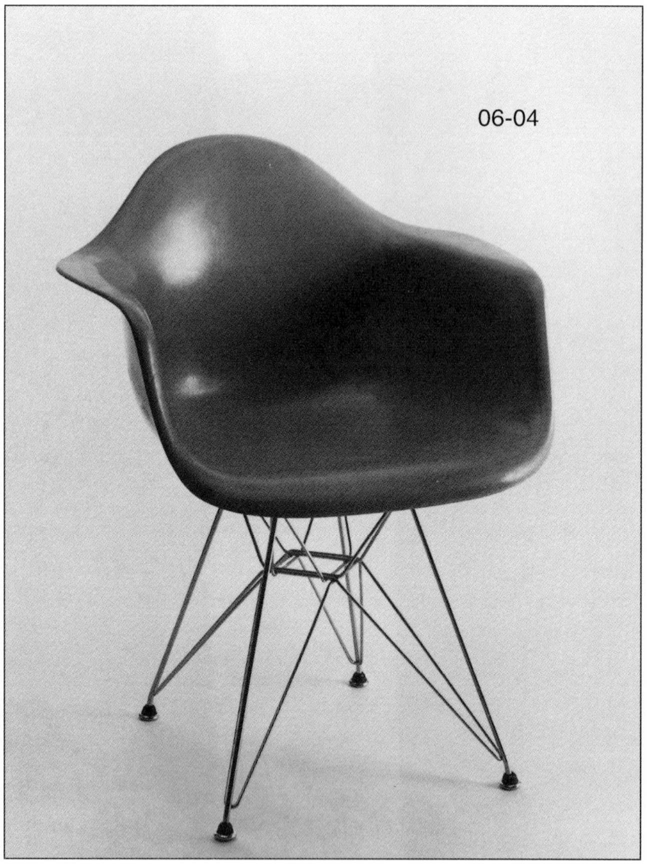
06-04

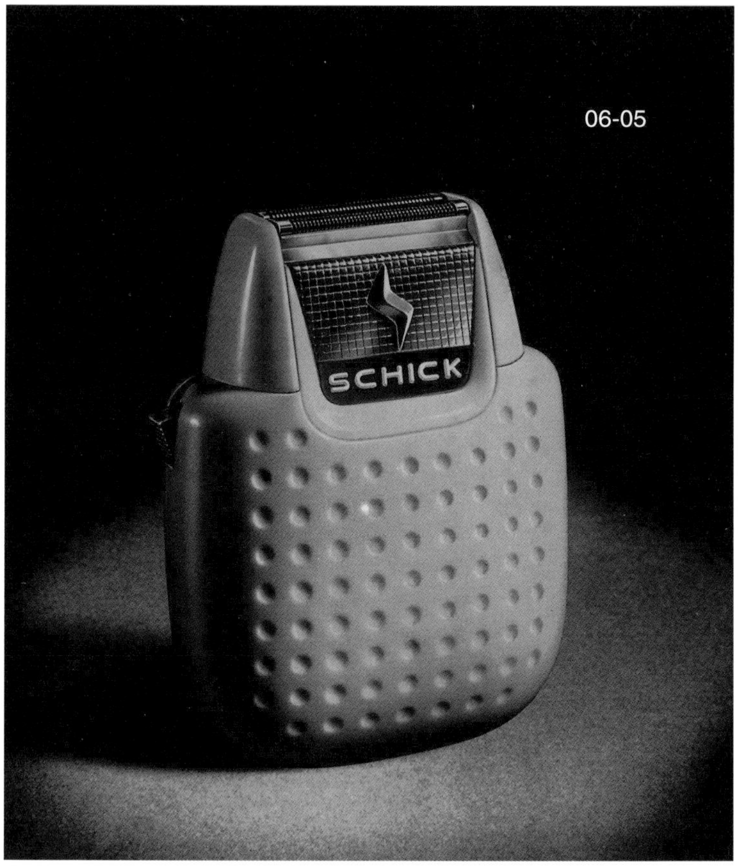

06-05

06-06

1951

The American Designers Institute (ADI) (see 1938) changes its name to the Industrial Designers Institute (IDI), and joins with the Chicago Society of Industrial Designers (CSID). The first president of the new organization is Paul MacAlister. IDI initiates the first national design awards, which will continue through 1965, when IDI will merge with other groups to form the Industrial Designers Society of America (IDSA) (see 1965). Awards include the Schick "20" shaver (see illustration) and the Eames chair (see illustration).

Design Research Inc. (D/R), a retailer of well-designed, mostly imported products, is started in Cambridge, Massachusetts, by Benjamin Thompson, a partner of TAC (see 1946), because it is difficult to find products of good design quality in U.S. retail shops. The first U.S. store for contemporary furnishings, D/R will soon open branches in New York and San Francisco. In 1969, Thompson will design its headquarters building in Cambridge, which in 2003 will receive the American Institute of Architect's coveted 25-year Award. D/R will survive until 1978.

The first design conference in Aspen, Colorado is initiated by Walter J. Paepcke, head of the Container Corporation of America (CCA), at the urging of CCA's consultant Herbert Bayer and CCA's Director of Design Egbert Jacobson (see 1936). It will become a major annual design event.

The *Luxo* drafting lamp, produced by Danish manufacturer Jacob Jacobsen (1902-1997), is introduced to the U.S. Jacobsen bought the *Anglepoise* lamp patent rights from Herbert Terry & Sons (see 1932), and founded Jac Jacobsen AS, later named Luxo. The *Luxo* light will sell in great numbers for more than 50 years.

The Museum of Modern Art holds an exhibition entitled, *8 Automobiles,* which were chosen "as works of art." Exhibited are the 1930 Mercedes-Benz Model SS, the 1936 Cord, the 1937 Talbot, the 1939 Bentley, the Lincoln *Continental* (see 1940), the Cisitalia, the Jeep (see 1941), and the MG Model TC.

06-04. This molded fiberglass-reinforced polyester, single-piece, shell-shaped chair (DAX), designed in 1948 by Charles Eames, is introduced by the Herman Miller Furniture Company. It is the first successfully mass-produced plastic chair and will be made through 1995 in many variations for dining, desk, lounge and rocking chairs that include wire bases known as the "Eiffel Tower." See 1953.

06-05. Schick introduces this Model "20" electric shaver designed by Carl Otto in collaboration with Norman Gray, chief engineer at Schick. The design reduces bulk and length by rearranging the internal components from the traditional narrow "in-line" configuration to a wider, but shorter one that is easier to hold. The design will become the new typeform for shavers.

06-06. General Motors exhibits *LeSabre,* a concept convertible with tail fins, "jet" intake for a grille, and the first true, curved, wrap-around windshield, in development by the Libbey-Owens-Ford Glass Company since 1946. Designed by Homer LaGassey and named after the North American F-86 *Sabre* jet fighter recently put into service in the Korean War, it will set GM styling trends for many years. The name *LeSabre* will be adopted by Buick in 1961 and will still be in use in 1995.

06-07

06-09

06-08

06-10

1952

Edgar Kaufmann Jr. of the Museum of Modern Art announces an exhibition, *Olivetti: Design in Industry.* This exhibit marks the first time a European company has been invited to show its products and graphic work in the U.S. The exhibition, curated by Mildred Constanine, will influence U.S. corporations, particularly IBM, in their awareness of design's strategic business value. See 1911,1956.

Walter P. Margulies, Frank Gianninoto, Egmont Arens, Alan Berni, Karl Fink, Robert Goldberg, Robert Gruen, Benjamin Koodin, Harry Lapow, James Nash, Robert Neubauer, and Gerald Stahl form the Package Designers Council (PDC). James Nash is named as President.

Raymond Loewy establishes an office in Paris as a hobby. Harold Barnett manages the new Compagnie de l'Esthetique Industrielle (CIE). It will become a serious business in 1957, when British Petroleum (BP) becomes a client. Barnett will remain until 1960. Douglas Kelley will succeed him and head the staff of 60 until 1966.

Pattern Highlight, a line of stainless steel flatware, designed by Russel Wright in 1950 for John Hull Cutlers Corporation, is first produced. The first US pattern of extreme modern design, Wright considers it his best, and calls it the "Pinch" design.

Alexander Kostellow creates an Experimental Design Laboratory at Pratt Institute, which provides office space adjacent to student drafting and shop areas for company designers to do experimental work and instruct students. Participating companies include General Motors, Shell Oil, and Owens-Corning Fiberglas Corporation. After Kostellow's death in 1954, Robert A. Kölli will succeed him at the lab as well as the Industrial Design Department at Pratt.

National design awards by the Industrial Designers Institute (IDI) include *Swingline* juvenile furniture by Henry P. Glass for the Fleetwood Furniture Corporation, and the General Electric *A-1 All Purpose Fan* (see illustration).

06-11

06-07. Don DeFano, Richard Latham, and Franz Wagner of Raymond Loewy Associates design this Model 1500 *Flight* bathroom scale. It will be introduced by the Borg-Erickson Company next year at $15.00, and will be selected by *Fortune* magazine as one of the top 500 designs of all time.

06-08. Thomas A. Edison, Inc. introduces this Edison portable *Voicewriter* Dictaphone (see IDI awards 1953), designed by Carl Otto.

06-09. General Electric introduces this *A-1 All Purpose Fan*, designed by Donald L. McFarland of General Electric Company.

06-10. Steel wire *Diamond* chairs, designed by Harry Bertoia, in low and high-back styles, are launched by Knoll Associates. They will become known later as *Bertoia Chairs.*

06-11. *Wedge-Lock,* cutlery with the first scientifically designed, ergonomic handle shapes, is introduced by Cutco Products of Olean, New York. The design was begun in 1941 by Thomas B. Lamb and patented in 1945. Based on Lamb's extensive study of hand mechanics and finger strength, the cutlery is sold through in-home demonstration by appointment only, and will continue to be sold this way into the next century. In 1954, Lamb's handle designs will be featured in an exhibition at the Museum of Modern Art and in a *Fortune* magazine article.

06-12

06-13

06-14

06-15

1953

The Industrial Design Institute (IDI) announces national design awards for the Edison *Voicewriter* (see illustration 1952), and the Servel *Wonderbar* portable refrigerette (see illustration).

Donald E. Dailey is named vice president of product planning for Servel, Inc. With the exception of Harley Earl, Dailey is the first industrial designer at this corporate level, and initiates the term and practice of "product planning."

Northern Industrial Chemical Company introduces *Residential*, the first plastic dinnerware for consumer use, designed by Russel Wright. Designed last year in *Melmac*, a melamine plastic, *and* available in mix-and-match colors, it will receive *Good Design* awards by the Museum of Modern Art this year and next. Acclaimed for its "organic" forms, the line will be highly successful (sales in the millions), and will prompt other manufacturers to enter the new plastic dinnerware market.

06-12. Servel, Inc, the sole manufacturer of gas absorption refrigerators since 1927, introduces the *Wonderbar*, a compact portable refrigerator made almost completely of Bakelite plastic. It is designed by industrial designer and Servel vice president of product planning, Donald Dailey. The *Wonderbar* is created to make refrigerators acceptable in rooms other than the kitchen. Servel also introduces the first automatic icemaker in its full-sized refrigerator, styled by Dailey and his staff .

06-13. Brunswick-Balke-Collender Company introduces this classroom school furniture designed by Dave Chapman. See IDI awards, 1954.

06-14. General Motors (GM) introduces the Chevrolet *Corvette,* the first U.S. sports car produced in high numbers, at its Motorama in New York. Designed by Harley Earl, engineer Edward N. Cole (1909-1977), and GM styling staff including Robert McLean, Clair McKeegan, and Larry Shinoda, it is the first car to have an all-fiberglass body. Although designed as a show car, it is an instant sensation and is rushed into production. It will soon become a classic and still attract purchasers at the end of the century.

06-15. Studebaker Corporation introduces this *Starliner* hardtop convertible in its *Commander* and *Champion* lines. Robert E. Bourke, assisted by Randall D. Faurot and Holden Koto, began designing the car in 1951. Bourke now heads the Raymond Loewy Associates group in South Bend, Indiana. The *Starliner* will be acclaimed by the Museum of Modern Art not as a design, but as "a work of art," and both models will be later known as "Loewy Coupes," or "Bourke Coupes." Unfortunately, sales will not be strong.

06-16. The 1953 *Good Design* exhibition (see 1950) at the Merchandise Mart in Chicago is designed by Alexander Girard and features dramatic white/black lighting. The exhibition includes this three-legged leather and steel "T-Chair" designed by William Katavolos, Douglas F. Kelley, and Ross Littell, 1949 Pratt classmates who cooperated with Laverne Inc. to produce a line named *The New Furniture.* The chair will become part of the Museum of Modern Art's permanent collection and will be produced continuously in Italy, after 1988, under the Cadsana name.

06-17. Herman Miller introduces this new series of DSX fiberglass shell chairs designed by Charles Eames, in forms that are smaller and simpler than earlier versions (see 1951).

06-16

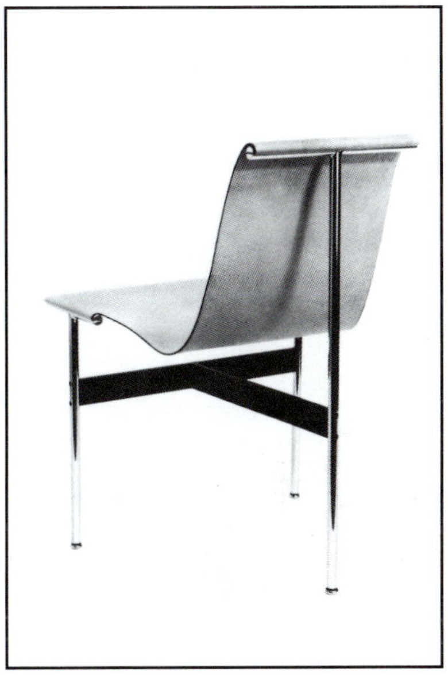

06-17

06-18

06-19

06-20

06-21

1954

The Tenth *Triennale* is held in Milan with the theme of *Industrial Aesthetics.* Begun as the *Biennale* (see 1923), the exhibition became La Triennale di Milano in 1931. The Society of Industrial Designers (SID) (see 1944) president, Peter Müller-Munk, and members send products to Milan for an exhibit of U.S. industrial design that includes a Millers Falls electric drill No. 1814 designed by L. Garth Huxtable, and two Buckminster Fuller geodesic domes. See 1957, 1959, and 1967.

The *Compasso d'Oro* (Golden Compasses) award for good design is established by La Rinascente department store in Milan. The Italian Associazione per il Disegno Industriale (ADI) will administer the awards starting in 1957.

A *Good Design* exhibition is held at the Merchandise Mart in Chicago. Museum of Modern Art jurors, headed by Edgar Kaufman Jr., selected 100 items from the previous five years for the retrospective exhibition of design excellence. See 1950.

The Industrial Designers Institute (IDI) presents annual design awards (see 1951) to Dave Chapman for school furniture (see illustration 1953), and to Raymond Loewy Associates for the Borg-Erickson bathroom scale (see illustration 1952)

Industrial Design, the first U.S. industrial design magazine publishes its debut issue. Founded by Jane Fiske Mitarachi (later Jane Thompson), the magazine is co-edited by Mitarachi and Deborah Allen, and designed by Alvin Lustig.

Dansk International Designs, founded by Danish designer Jens Quistgaard and U.S. businessman Ted Nierenberg (b. 1919), begins production of household products that interpret the beauty of wood with modern forms.

Giulio Natta of Italy creates a new polypropylene plastic with the exceptional quality of flexing repeatedly without breaking. Based on work by Karl Ziegler of Germany, it will be produced commercially in 1957, and the Nobel Prize for chemistry will be awarded jointly to both men in 1963.

The first radio using transistors, the *Regency TR-1,* is introduced at $49.95 just in time for Christmas sales. Created in six weeks, an unbelievably short time, the radio is a collaboration between Texas Instruments in Houston, Industrial Development Engineering Associates in Indianapolis, and the designers, Painter, Teague, and Petertil in Chicago.

The L.M. Ericsson Company introduces the *Ericofon 700,* a compact one-piece plastic desk telephone that includes an earpiece, dial, mouthpiece, and switch in one unit, designed by Swedish industrial designer Gosta Thames. The organic design with the dial on the base bottom was in development since 1940 by Ralf Lysell and Hugo Blomberg; it is hailed by the design journal *Form* as "technical sculpture."

06-18. *Design 1,* stainless steel flatware designed by U.S. designer Don Wallance, is manufactured by Pott GmbH, Germany for H.E. Lauffer, New York, and is introduced in the U.S.

06-19. This General Electric (GE) kitchen concept including a wall-hung refrigerator is featured in *Life* magazine. Bob Munz, Bill Saenger, and Harrison Linger designed the kitchen under Arthur BecVar of GE.

06-20. Procter & Gamble introduces *Gleem* toothpaste, with this packaging by Donald Deskey Associates.

06-21. Bissell Carpet Sweeper Company introduces this *Sweepmaster,* designed by James G. Balmer, Jr., and Frederick W. Hertzler, and engineered by Carl B. Denny, all of Harley Earl Associates, Inc. (see 1945). See 1955 for award.

06-22

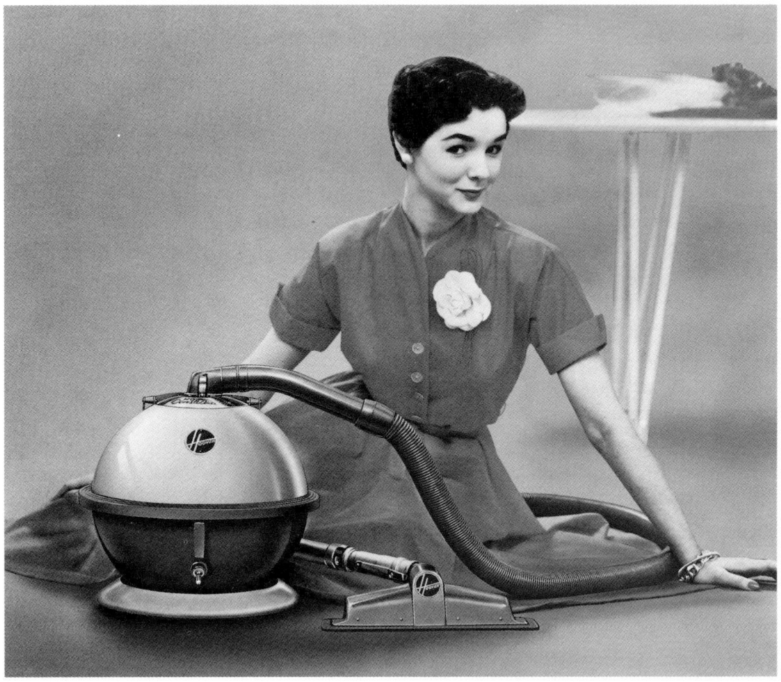

06-23

06-24

1955

The Hochscule für Gestaltung (HfG) at Ulm, Germany opens, declaring itself the "New Bauhaus." The school was established to foster national economic recovery, and was funded by the German government and the American High Commission for Germany. Max Bill is named director, with Tomas Maldonado as deputy. The focus on product design will be abandoned in 1956 in favor of theory and information problem solving. Bill will resign in 1957, Maldonado in 1966, and, in 1968, while directed by Herbert Ohl, the school will be dissolved by the legislature of Baden-Württenburg.

Max Braun's sons Artur and Erwin reorganize Braun AG in Germany. Dr. Fritz Eichler directs a new design program in collaboration with the new Hochscule für Gestaltung (HfG) at Ulm, and engages Hans Gugelot (1920-1965) and Otl Aicher (1922-1991), as well as Herbert Hirche from the Stuttgart Academy of Arts, and 23-year-old Dieter Rams. In 1956 the new department will establish the most consistent corporate and product aesthetic style to date, using white spare forms in the classic Bauhaus tradition. By 1957, Rams will head the program, which will achieve global recognition as a corporate model of good design. The Gillette Company will purchase Braun in 1967.

The Industrial Design Institute (IDI) presents design awards for the Bissell *Sweepmaster* (see illustration 1954); to Randall D. Faurot for the design of the *Lectracar Duo* self-propelled vehicle; and Richard Montmeat of General Electric industrial design staff for its *660 Convertible* clock radio. IDI's new national headquarters open in New York. The Society of Industrial Designers (SID; see 1944) changes its name to the American Society of Industrial Designers (ASID).

Sony of Japan, founded in 1946 as the Tokyo Kabushikakaika Company, introduces the TR 55 transistor radio, using U.S. transistor patents and the Sony name for the first time. It will become the first transistor radio success in the market.

General Electric (GE) offers many of its major appliances in "mix-or-match" bright colors. Color appears on Westinghouse's mixers in yellow, turquoise, and pink, and on Presto's iron handles and coffee makers. GE's "Woodtone Brown," later known as "Coppertone," will soon be a standard option color on all major appliances. See 1966,1977, 1986.

The Museum of Modern Art's *Good Design* program (see 1950) ends with the resignation of Edgar Kaufmann Jr. The last exhibition shares billing with "Good Companions," rooms of traditional furniture. Arthur Drexler will replace Kaufmann next year as director of the Museum's Department of Architecture and Design.

06-22. The Hoover Company introduces its Model 82 *Constellation* vacuum cleaner, a sphere that glides on an air cushion of its own exhaust. Designed by Henry Dreyfuss Associates, it is known as "The Cleaner that Walks on Air."

06-23. Ford's new *Thunderbird* is designed by Bill Burnett, William F. Boyer, and Franklin Quick Hershey. It will quickly become a classic as the "T" Bird. George Walker is appointed Ford's vice president of styling.

06-24. The Frank G. Hough Company introduces this Model HH *Payloader* power shovel, designed by Jon W. Hauser. See 1956 for IDI award.

06-25

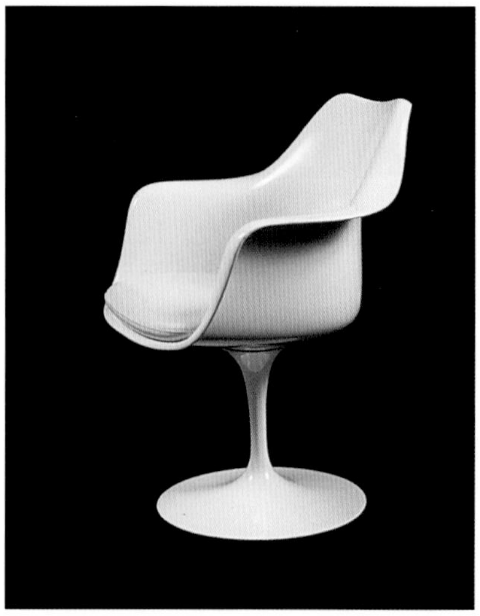

06-26

06-28

1956

Thomas J. Watson Jr., president of IBM, engages Eliot Noyes, a glider pilot friend, as consultant design director to improve the visual quality of IBM products, graphics, exhibitions, interiors, packaging, and architecture in order to compete esthetically with Olivetti (see 1952). Noyes engages additional consultants, including Paul Rand, Charles Eames, and Edgar Kaufmann, Jr. The enterprise will become the earliest design realization of a total corporate "house style" program.

The Industrial Designers Institute presents its annual awards to Jon W. Hauser for the *Payloader* (see illustration 1955), George Walker for the Lincoln *Premier Coupe*, and William E. Clements for a *Thermameter* for the Medical Research Institute, Inc.

The Council of Industrial Design (see 1944) opens the *Design Centre* in central London, a national information and exhibition facility for industrial design.

The Human Factors Society (HFS) is organized to "increase and diffuse the knowledge of man in relation to machines and his environment." It will later be named the Human Factors and Ergonomics Society (HFES).

The Herman Miller Furniture Company introduces a lounge chair and ottoman designed by Charles Eames that are molded rosewood plywood shell structures padded with leather cushions. Originally designed as a birthday gift for film director Billy Wilder, the chair set is quite expensive. Yet the design will become a classic, being placed in the Museum of Modern Art's permanent collection, and winning a *Compasso d'Oro* at the *XII Triennale* in Milan in 1960. Herman Miller also introduces the *Coconut Chair and* the *Marshmallow Chair*, both designed by George Nelson.

RCA introduces a "personal" portable color television set, the industry's first eight-and-one-half-inch screen, which weighs only 22 pounds, and will win honors at the 1957 *Triennale* in Milan. Henry Rundle, manager of product development, designed the television set at the RCA-TV Division, Camden, New Jersey.

06-25. Knoll Associates introduces this *Pedestal Chair* and related table and stools in molded fiberglass, a design Eero Saarinen began in 1953 with assistance from Don Petit of Knoll's Design Development Group. The graceful white organic shapes are supported by a single stem flowing from a circular base, the stated design objective being to reduce the profusion of legs in a typical room setting. It will be Saarinen's last major furniture design.

06-26. Volkswagen introduces the *Karmann Ghia* coupe in the U.S., an alternate body style for its standard *Beetle,* designed by Carrozzeria Ghia in Turin, Italy.

06-27. The wall-mounted telephone common before 1900, is re-introduced by Bell Telephone Laboratories. Designed by Henry Dreyfuss Associates, the phone is intended as a companion to the desktop model "500" (see 1949), and will become very popular in kitchens where counter space is at a premium.

06-28. The Dripcut Starline Company introduces this line of glass and chrome table dispensers for sugar, cream, salt, and pepper, designed by Henry Keck. More than 13 million sets will be sold and The Traex Company in Dane, Wisconsin, will still manufacture them in 2001.

06-27

06-29

06-30

06-31

06-32

1957

The International Council of Societies of Industrial Design (ICSID) is founded in London. It is comprised of 23 professional and promotional design organizations from 17 countries. Peter Müller-Munk of the American Society of Industrial Designers (ASID) (see 1955) is elected its first president.

Because of Walter Dorwin Teague's personal appeal to the U.S. Information Agency (USIA), for the first time in 35 years, the U.S. government funds an exhibit for this year's Italian *Triennale* (see 1954). Paul McCobb designs the U.S. exhibit to house 200 U.S. products in a geodesic dome by R. Buckminster Fuller. L. Garth Huxtable again (see 1954) wins an award for his design of the Millers Falls *Plane-R-File*.

The book *Hidden Persuaders,* by Vance Packard, claims that advertisers manipulate customer wants and needs with motivational research. Designers are implicated by association, especially Brooks Stevens, to whom Packard refers as "the crown prince of 'planned obsolescence.'"

Industrial Designer's Institute design awards are announced for the 1957 Chrysler cars (see illustration), IBM's RAMAC, and the *General Electric Kitchen Center* (see illustration).

Polycarbonate and other structural plastics are introduced. General Electric's *Lexan* was developed jointly with Bayer AG. These new materials will not be in full production until 1960.

The Industrial Design Education Association (IDEA) is founded, with Joseph Carriero (1920-1978), chairman of the industrial design department at the Philadelphia Museum School of Art, as its first president. Educators formed their own organization because neither professional society (IDI or ASID) accepts educators as full members.

Wham-O introduces a toy plastic disk called *Pluto Platter,* developed by Walter Frederick Morrison. Before 1951, it was called *Morrison's Flying Saucer.* Ancestors of such toys were pie tins made by the Frisbie bakery in Bridgeport, Connecticut, that sailed through the air upside down, and had been used since 1920 by Yale students. The toy will become known as the *Frisbee* in the 1960s, after Wham-O promotes it not as a toy, but as a sport.

General Electric introduces the first combination spray, steam, and dry iron, Model F61, designed by W.E. Moore, Paul O. Rawson, and Rudolph H. Koepf. It will set the design typeform for years.

06-29. This Braun Model KM 3/KM 31 *Kitchen Machine*, designed by Gerd Alfred Müller, is introduced. It will be the first Braun product to attract widespread international interest.

06-30. General Electric presents this new Kitchen Center designed by Arthur N. BecVar and Robert W. Blee. The grouping includes a combination washer/dryer, a *Disposall*-equipped sink, a dishwasher, and a range with oven—all under a single, seamless stainless steel countertop.

06-31. A complete new line of Chrysler Corporation cars is designed by Virgil Exner (in photo), Henry T. King, H.T. Bannister, Clifford C. Ross, Carl Reynolds, and Robert E. Bingman (1903-1967). Termed the "Forward Look," it is a big hit.

06-32. Monsanto's experimental fiberglass house called the *House of the Future* includes prefabricated fiberglass bathrooms designed by Henry Dreyfuss Associates. The house includes this *Kitchen of the Future* developed by the Kelvinator Division of American Motors Corporation under Randall D. Faurot (shown in photo) by Don Khaki, Harold Smith, and Douglas Wolff.

06-33

06-34

06-35

06-36

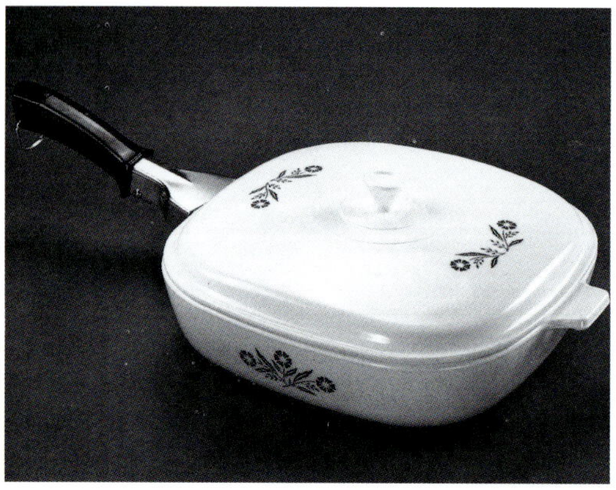

06-37

06-38

1958

Industrial Designers Institute (IDI) awards go to 3M's *Secretary* (see illustration), the *Pedestal Group* (see illustration 1956) by Eero Saarinen, and the AMI *Portable Washer* by Melvin H. Boldt.

The Aluminum Company of America (ALCOA) initiates its *Forecast* series of advertisements promoting designs in aluminum, featuring concepts by well-known designers including the *Solar Machine*, a sun-powered "do-nothing machine" designed by Charles Eames.

The Adolph Coors Company develops lightweight aluminum cans, produced by Kaiser Aluminum Company. In 1961 Kaiser will develop aluminum "soft-tops" and, by 1963, mass-manufacture of these cans will create an enormous environmental problem.

The *Hula Hoop,* a plastic version of a recent Australian bamboo toy, is introduced by Wham-O, whose owners are Richard Knerr and Arthur K. "Spud" Melin (1925-2002). Similar toys were known from ancient Egypt but the plastic hoop quickly becomes a national fad.

Ford Motor Company introduces the *Edsel,* named after Henry's son. Designed by chief Edsel stylist Roy A. Brown under vice president of styling George Walker with exhaustive market research by David Wallace, it is a huge failure, and losses of $350 million will force cancellation of the brand in 1959.

Philco breaks away from square television sets with its new *Predicta.* Designed by Philco staff under design manager Herbert Gosweiler, the tube is a separate shape that sits on top of the cabinet.

06-33 & 34. U.S. Steel introduces this new logotype and an industry-wide "Steelmark" designed by Lippincott & Margulies. Even after the "Steelmark" falls from usage in the 1980s; it will find eternal life as the trademark of the Pittsburgh Steelers. See 1945,1962.

06-35. Minnesota Mining and Manufacturing Company (3M) introduces *The Secretary* copying machine, designed by Samuel M. Highberger of Harley Earl, Inc. for Thermo-Fax.

06-36. Nathan George Horwitt designs a wristwatch that has a black face without numerals and a white disk marking the 12 o-clock position. Zenith Movado produces the watch without credit or compensation to Horwitt, and it will become a classic example of design piracy. It will be placed in the permanent collection of the Museum of Modern Art in 1960, and become known as "The Museum Watch." In 1975, Movado will settle with Horwitt for $29,000 and by 1997 will be heavily promoting Horwitt and the design.

06-37. *CorningWare* cookware, a freezer-to-oven-to-table line made of *Pyroceram®* material, is introduced by Corning Glass Works. The pieces feature the first blue *Cornflower* decoration designed by Corning staff, which will become the trademark of Corning consumer products for three decades.

06-38. Boeing's 707, the first commercial "jumbo jet," goes into service, with interiors and exterior color scheme by Frank del Guidice, of Walter Dorwin Teague Associates. One of the planes will become *Air Force 1* for President Kennedy in 1963. The 707 will be produced until 1991. See 1969.

06-39. Underwood Typewriter Company introduces this *Documentor* electric typewriter designed by Raymond Spilman, which will set a new typeform for electrics. More than half of all typewriters sold today are electric.

06-39

06-40

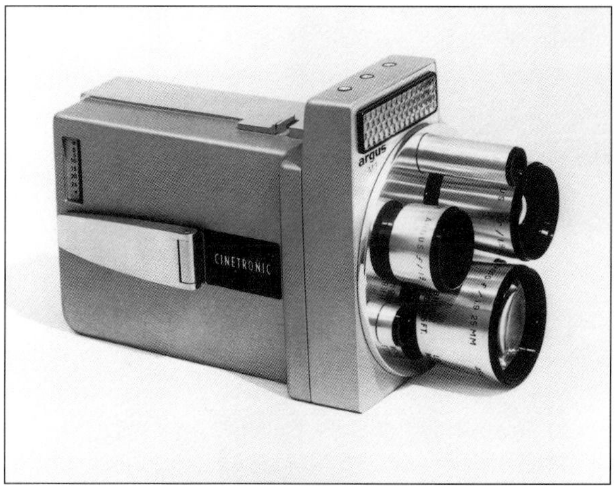

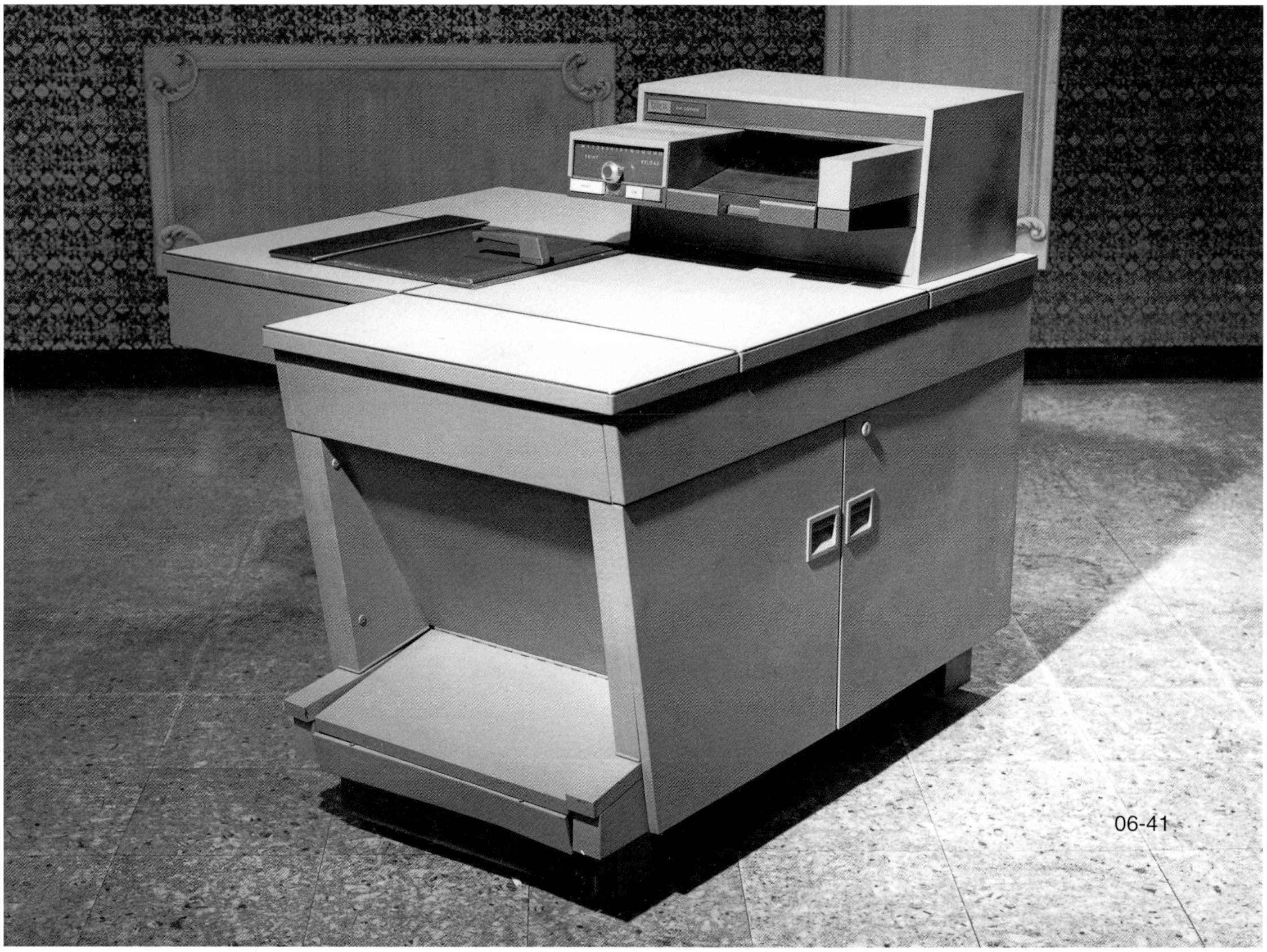

06-41

06-42

1959

A U.S. Information Agency (USIA) exhibit, *American National Exhibition of American Culture,* designed by George Nelson, opens in Moscow. It includes presentations by Charles Eames, a geodesic dome by Buckminster Fuller, and displays that feature modern kitchens by RCA/Whirlpool, General Electric, and a model suburban home kitchen designed by Raymond Loewy Associates, where the famous "Kitchen Debate" between Russian Premier Krushchev and Vice President Nixon takes place. See 1966, 1967.

An article in *Fortune* magazine, *Old Business, This Industrial Design*, surveys a total of 300 design organizations, 180 consulting offices, and 120 corporations, of which 180 are members of ASID or IDI. Independent designers see the growth of internal corporate staffs as "a threat to creative design", and refer to them as "captive designers."

ALCOA initiates Student Design Awards to students of five participating schools, including a powered skateboard by Louis Skeeter of IIT. The program will continue through 1971. ALCOA also presents its first Industrial Design Awards to Charles Eames, Peter Muller-Munk, and Fred. W. Priess.

Mattel, Inc. introduces *Barbie*, a doll that represents a teenager, at the New York Toy Fair. Mattel co-founders Ruth (1917-2002) and Elliot Handler named the doll after their daughter and based the design on a sophisticated West German doll, *Lilli. Barbie* will become the best-selling doll in history, selling more than one billion by 1998.

Morris Motors of England (BMC) introduces a compact, front-wheel drive *Mini Minor*, with a transverse engine in order to provide adequate passenger space. Alec Issigonis (1906-1988), who will become Morris's chief engineer in 1961, designed it. Nearly 5.5 million *Minis* will be produced, and they will be sold in the U.S. until safety and emissions standards close the market. A modified version named the Mini *Cooper* will be launched in the U.S. in 2001 by BMW.

06-40. Industrial Designers Institute (IDI) awards this year include this *Cinetronic Electric Eye* 8mm movie camera by Ray Grosso, Fred W. Hertzler, and Dominic A. Saporito, of Harley Earl, Inc. for Argus Cameras, a division of Sylvania Electric Products, Inc.

06-41. The first automatic office copier to make photocopies on plain paper, the *914,* is introduced by Haloid Xerox. The company will become the Xerox Corporation in 1961, and its name will become a synonym for "copy." The *914* was designed by James G. Balmer, Jr. of Armstrong/Balmer & Associates, Inc. in collaboration with Don Shepardson, John Rutkus, and Hal Bogdenoff of Xerox, and 200,000 will be produced by 1976. See 1963.

06-42. Bell Telephone Laboratories introduce the *Princess* telephone, with a hand/mouthpiece that spans the dial and nests compactly on the base. Designed by Henry Dreyfuss Associates, the phone's petite size aims to appeal to teenage girls. Its illuminated dial will encourage use as a phone for the bedroom.

06-43. Donald E. Dailey of Don Dailey Associates, Inc. designs this twin wheel caster for Faultless Caster Company which eliminates the unsightly "yoke" for the axle by hiding it between two wheels. After it is Introduced in 1963, it will be copied and will revolutionize the furniture industry.

06-43

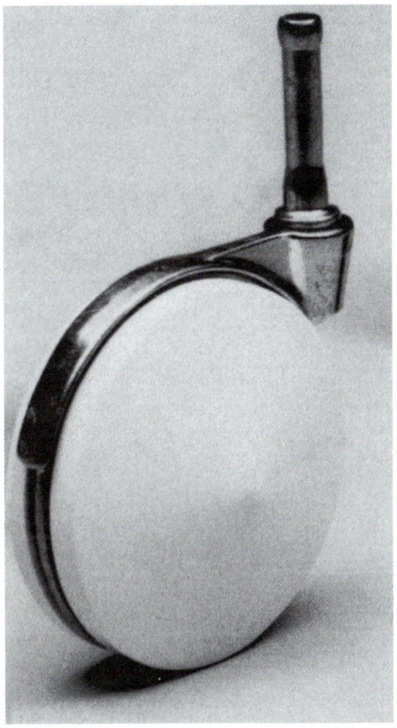

Chapter 7

1960-1969 Responsive Design

During the 1960s, an international visual language joins U.S., Scandinavia, and Italian design. U.S. consumer goods manufacturers have embraced the new contemporary design trends. Many U.S. furniture, home accessories, home equipment, and housewares manufacturers organize internal design departments. New manufacturers of computers, office machines, televisions, and electronic products apply the spare, trim aesthetic that typifies this consistent international style, ignoring Detroit's flamboyance. Metal materials suppliers, challenged by the growing plastics industry, promote good design in products using their materials.

As corporations refine their management structures in order to deliver products more appealing to target audiences, they become more competitive. Businesses find that marketing, design, and packaging are essential to communicate brand and product values, and to merchandise these products domestically and internationally,

Consumers, no longer willing to accept poor design quality in products, are attracted to the value of imported products in automobiles, home equipment, or cookware. They become suspicious of market manipulation and unnecessary product features. New choices in the marketplace, represented in the U.S by stores like Crate & Barrel and in Great Britain by Habitat, offer affordable imported products for sophisticated, young tastes.

Perhaps most significantly, many young consumers have never experienced recessions or material deprivation. As the "baby boomers" reach adulthood, they exert powerful influence upon all consumer markets, to say nothing about their impact on U.S. culture, mores, and perception of value. This new post-war generation wants much more from design than style. The youth market seeks substance, value, honesty, and responsibility.

Some designers question how to respond to these non-traditional consumer demands, which differ so dramatically from the post-war materialistic and acquisitive culture. International tensions, global competition, the Cold War, the technological embarrassment of Soviet space achievement, civil rights strife, the assassinations of President John Kennedy; Robert Kennedy, and Martin Luther King, and the Vietnam War contribute to the social turmoil. Young people, deeply affected by these events, are quick to rebel against the establishment and search for alternatives. Activists turn the attention of business and design to concerns such as consumer protection, auto safety, environmental risk, and corporate greed.

The design world mourns the death of modern design pioneers Walter Gropius, Mies van der Rohe, Henry Dreyfuss, Peter Müller-Munk, Egmont Arens, and Harley Earl. With these and other founding fathers now gone, new leaders and directions are needed. Young designers view the design establishment as out of touch with the social challenges of the day even when the three industrial design organizations combine into a single voice to seek joint purposes. The design community faces a generational divide, both organizationally and professionally.

The validity of both International Style and General Motors style is questioned. Leading architects explore "Postmodernism" and "Deconstructivism." The National Endowment for the Arts is established; it becomes an agent for change in design disciplines. At the end of the decade, the Apollo 11 landing on the moon and development of the supersonic *Concorde* refresh imaginations with technological optimism, but young designers are seeking a purpose beyond the traditional product appearance role of the profession's pioneers.

07-01

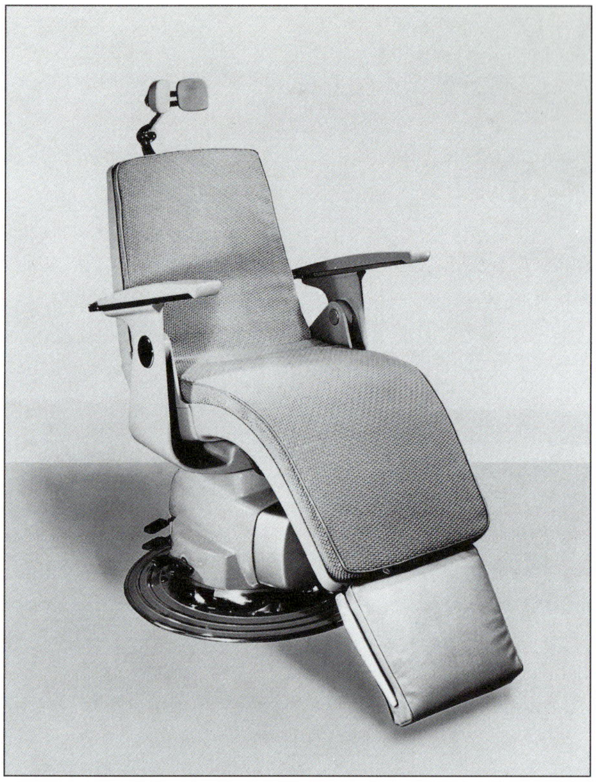

07-02

07-03

1960

Westinghouse, hoping to increase its consumer products market share against giants General Electric and Whirlpool through design, hires Eliot Noyes to dramatize the company's image, which has become "non-descript and unsightly." Noyes engages graphic designer Paul Rand, and recommends strengthening the internal design organizations. See 1964.

A Bauhaus archive is established in Darmstadt, Germany, on the initiative of art historian Hans M. Wingler, who is named director. In 1962, Wingler will publish *Bauhaus,* a massive volume of documents and photographs, and in 1976 the German Democratic Republic will restore the Bauhaus buildings in Dessau to their former glory.

The Institute of International Education, in collaboration with the International Council of Societies of Industrial Design (see 1957), initiates the Kaufmann Award of $20,000 to recognize design excellence. The first award goes to Charles and Ray Eames "for the practice of design." See 1964.

Industrial Designers Institute (IDI) annual awards go to Chevrolet's *Corvair* (see illustration), Ritter's *Euphorian* dental chair by Walter Dorwin Teague Associates (see illustration), and to Dave Chapman, Kim Yamasaki, and Frank Carioti for their research report and design recommendations to the Educational Facilities Laboratories, Inc., for U.S. school facilities (see furniture illustration 1953).

Thomas G. Hardie imports the first non-stick skillet made by T-Fal Corporation of France to the U.S. Engineer Marc Grégoire founded T-Fal in 1956 (as Tefal) after discovering a process to coat pans with polytetrafluoroethylene (PTFE). DuPont will soon counter with its *Teflon* coating as demand for non-stick products expands. See 1938.

The Whitney Library of Design publishes *Measure of Man,* by Henry Dreyfuss, an ergonomic data guide along with diagrams that incorporated body measurements compiled from military records by Alvin Tilley of the Dreyfuss office. It features "Joe" and "Josephine" (see 1943), life-sized anthropometric adult male and female figures, and will popularize the idea of fitting products to human scale.

Chevrolet's new compact car, the *Corvair,* designed by Ron Hill and GM Styling Staff, enters the market and is acclaimed by the design community (see IDI awards). It is plagued by an over-steering tendency that causes fatalities, and Ralph Nader later calls it "one of the nastiest-handling cars ever built." Nader's criticism of auto safety hazards will initiate government safety regulations in the auto industry.

07-04

07-01. This *Euphorian* dental chair, completed last year by the Ritter Company and designed by Walter Dorwin Teague, Jr., David H. Deland and Benjamin H. Stansbury of Walter Dorwin Teague Associates, reaches the market.

07-02. This new Polaroid *Electric-Eye 900* camera, designed by Walter Dorwin Teague Associates, debuts at $159.95.

07-03. Copco Incorporated is formed by Sam Farber (see 1990) to manufacture and distribute porcelain-enameled cast iron cookware designed by Michael Lax, and *Margrethe* mixing bowls designed by Bernadotte & Bjørn (see 1950). Copco will launch cookware shops in specialty, gift, and department stores. Although Lax will establish his independent design office in 1965, he will continue design for Copco for more than 20 years.

07-04. Paul Rand, working with Eliot Noyes, re-designs this new "Circle W" trademark for Westinghouse, and begins work on a corporate graphics design manual that will be completed in 1961.

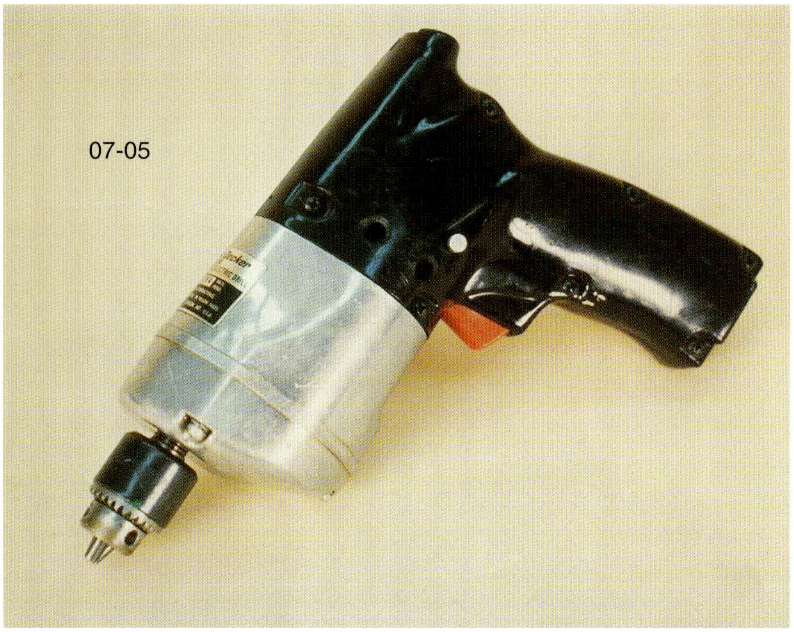
07-05

07-06

07-07

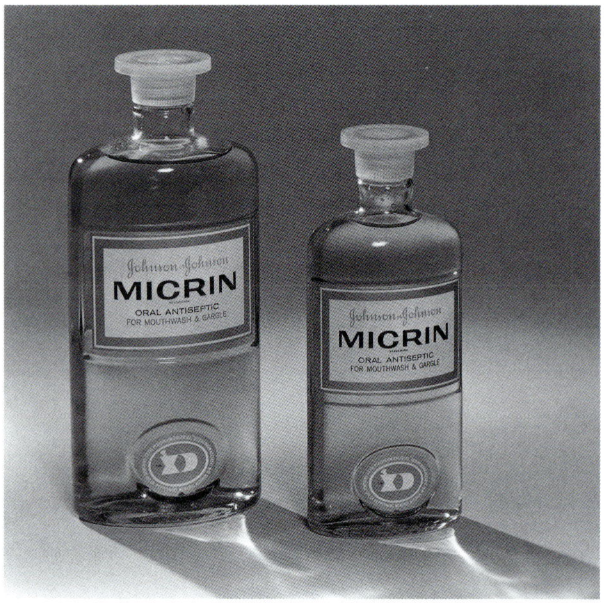

07-08

07-09

1961

The second $20,000 Kaufmann International Design Award goes to Walter Gropius for "achievement in design education."

The National Design Council is established in Canada by an Act of Parliament "to promote and expedite improvement of design in the products of Canadian industry." See 1976.

Industrial Designers Institute (IDI) presents its annual design awards to the Ford styling staff for the *Continental*; to Sundberg-Ferar for the Sears' *Tower Capri* typewriter; and to Arnold J. Copeland, Robert Baker, and Robert Y. Kimura of Visual Marketing, Inc. for its design of Educational Project Kits for General Electric Company.

Minnesota Mining and Manufacturing Company (3M) introduces the *Decor Dispenser* Model C-15, designed by Jean Reinecke. The swooping plastic desktop tape dispenser, in numerous colors and textures, will be in use through the end of the century. See 1940.

Pampers, the first successful, mass-produced disposable diapers, developed by chemical engineer Victor Mills at Proctor & Gamble, debuts. His granddaughter, Marion Donovan, inspired the idea by making prototypes of shower curtain material.

Briton Alex Moulton (b. 1920) introduces the Moulton bicycle, a city bike that uses a cross-frame design with small wheels and featuring an unusual seat height adjustment.

07-05. Black & Decker U.S. introduces this, the first cordless consumer drill using rechargeable nickel-cadmium batteries developed by General Electric for NASA's space program. Battery-operated products in the past have been perceived as a nuisance, because batteries often needed replacement when products were needed most. Black & Decker establishes an internal design department headed by Marshall B. Johnson.

07-06. The Radio Corporation of America (RCA) unveils its "Sets of the Seventies," futuristic style concepts by its Advanced Design Center under Tucker Madawick, manager of industrial design. John L. Burns, president of RCA, created the Center in 1960.

07-07. This horizontal *Toast-R-Oven,* Model T-93, with front pull-down door and drawer access and designed by staff designers Paul Rawson and Roger Funk, is introduced by General Electric. This *Toast-R-Oven* design is similar to an earlier 1956 Munsey toaster, but will propel this new product category into a best seller.

07-08. Johnson & Johnson Company introduces *Micrin,* a new mouthwash, in this bottle designed by Donald Deskey Associates.

07-09. IBM introduces this revolutionary *Selectric I* electric typewriter which replaces the standard typebars with a spherical "golf ball" printing element that moves while the carriage remains fixed. The ball elements are available in several fonts and are easily interchangeable. Eliot Noyes began designing the sculptured housing in 1959. By 1975 the *Selectric* will account for about 75 percent of the U.S. electric typewriter market.

07-10. The only amphibious passenger automobile ever to be mass-produced, the *Amphicar,* designed by Hans Trippel, debuts. Powered by a rear four-cylinder Triumph Herald engine producing 43 h.p., all are convertibles and are available in four color choices. The front wheels act as rudders in the water. Manufactured in Germany by the Amphicar Corporation, 90 percent of the cars sell, at $2800 each, in the U.S. About 3700 will be produced by 1968, when federal regulations will force the company from business.

07-10

CHRYSLER
CORPORATION

07-11

07-12

07-13

07-14

1962

The *Century 21* Exposition, *Man in the Space Age,* is held in Seattle, Washington, and is symbolized by a 600-foot tower named the "Space Needle."

The Olivetti Company wins the Kaufman International Design Award of $20,000 "for outstanding use of design."

Industrial Designers Institute presents awards for the *Michelob* beer bottle (see illustration), and to John Peter Ronvik of the Design Department of Sears, Roebuck & Company, for design of the *Sears Tower Varizoom* camera by Bell and Howell Company.

The Soviet Union establishes the All-Union Research Institute of Industrial Design, appointing Yuri Borisovich Soloviev as Director. Soloviev was born in 1920, trained in the arts, architecture, and engineering, and has been an industrial designer since 1943.

The Color Marketing Group is founded, an association of color design professionals who forecast color directions two years in advance. This year, white is the most popular color in housewares (29 percent), with wood tones second (20 percent).

Samsonite Corporation organizes a corporate industrial design subsidiary under the leadership of Robert K. Fujioka, formerly of Melvin Best Associates. The group is named Design West, Inc., and has the unusual corporate freedom to solicit and provide design services in the open market.

Moen Inc., introduces its 722 kitchen faucet with a single control lever designed by Anthony J. Cascarano for Mel Boldt & Associates. Alfred Moen (1917-2001) invented the single-knob faucet while a college student in 1937. His company sells more than 70 percent of all kitchen faucets in the U.S.

07-11. Lippincott & Margulies designs Chrysler's new corporate identity to include this pentagonal logotype.

07-12. Anheuser-Busch, Inc., introduces this new, sensuously shaped beer bottle for *Michelob* designed by Jerome Gould, of Gould & Associates Inc.

07-13. Studebaker Corporation, which changed its name from Studebaker Packard this year, introduces its revolutionary *Avanti* car designed by Raymond Loewy Associates. Loewy had last worked for Studebaker in 1955. In 1961, new Studebaker president Sherwood H. Egbert wanted to demonstrate that Studebaker was still an innovative company, so he called upon Loewy. To design the *Avanti*, Loewy retreated to his home in Palm Springs, California, with designers Bob Andrews, Tom Kellogg, and John W. Ebstein. The design will be an instant success, but Studebaker will fold in 1963. The *Avanti II*, however, will be built beginning in 1965 by the Avanti Motor Corporation, formed by Nathan Altman and Leo Newman, after they buy the Studebaker production line. Production will increase to 150 cars a year by 1976, the year of Altman's death. Avanti will be sold in 1982 to Stephen Blake, and again in 1986 to Mike Kelley and John Cafaro who will move it to Youngstown, Ohio, in 1987. It will change ownership again in 1988, before it seems to die in 1991.

07-14. Claire Samhammer of Melvin Best Associates designs the *Classic* attaché case introduced by Samsonite Corporation. With its flush latches and slim, chrome-trimmed, charcoal gray case, the briefcase will become a classic in the corporate army of "gray flannel suit" businessmen for years to come. See 1965.

07-15

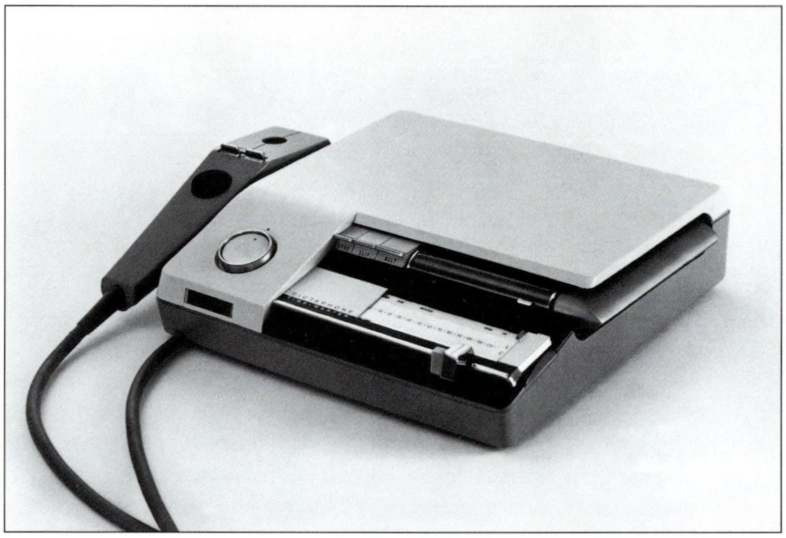

07-16

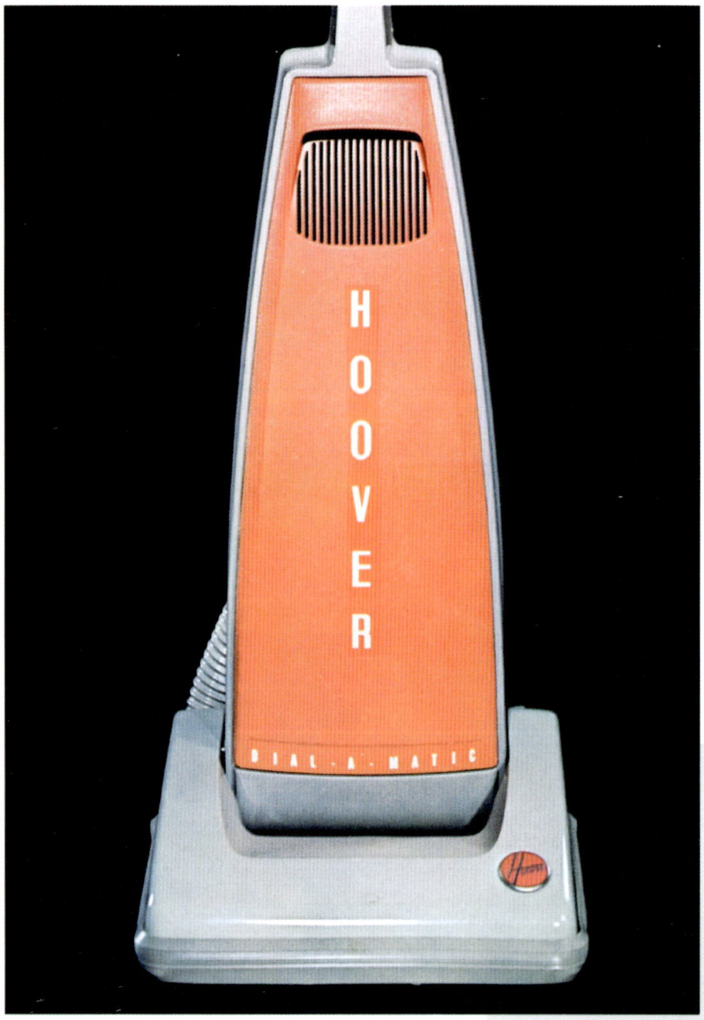

07-17

07-19

07-18

1963

Ermal Cleon Fraze, (d.1989) founder of the Dayton Reliable Tool & Manufacturing Company, patents an "ornamental design of a closure with a tear strip opener" for soda cans he has worked on since 1959. There was an earlier "Pull-top" design introduced by Can-Top Machinery Corporation in 1961, and ALCOA introduced a "zip-top, flip-top, pop-top, tear-out top" in November 1962. But Fraze's design, with the rivet attaching the "tab" formed from the lid itself, will become the standard in the industry, when he sells exclusive rights to ALCOA. See 1976.

The International Council of Societies of Industrial Design (ICSID—see 1957) convenes its biennial Congress in Paris. It holds an industrial design exhibition organized by Alfons Bach of IDI (see 1951), at the Louvre that includes Studebaker's *Avanti* (see illustration 1962), IBM's *Selectric* typewriter (see illustration 1961), Lincoln's *Continental*, and a range of U.S. products.

The International Council of Graphic Design Associations (ICOGRADA) is founded; it will hold a biennial congress.

The Industrial Designers Institute (IDI) presents annual awards for design of the Dictaphone *Time-Master* (see illustration), the GE *Square Pulse Bonder,* and the *Fin Light* by Douglas Deeds for Peter Pepper Products. Certificates of Design Merit include the Kodak *Instamatic* Camera (see illustration), the Pontiac *Grand Prix,* and the Caloric *Second Oven.*

The American Iron and Steel institute (founded 1870) initiates its *Design in Steel* awards program for Canadian, Mexican, and U.S. designers. Ten winners, ranging from bridges and architecture to the *Pal* razor, are announced in April. The annual awards will continue for about 10 years.

The *Astro Lamp*, a mesmerizing, glowing, multicolored oil and wax ooze in a hurricane lamp-like glass shape, is introduced. Designed and produced by Edward Craven Walker (1918-2000), a British Royal Air Force Squadron Leader, it is based on an earlier patent by David George Smith. The lamp will become known as the *Lava Lamp*. By the mid 1970s it will become an icon of the psychedelic age, selling seven million per year. Though the fad will soon pass, it will never go out of production, and in 1989 will be reintroduced in England, and become a staple item for the bedroom decor of teenagers into the next century.

07-15. Kodak introduces this ready-to-shoot *Instamatic 100* camera, designed by Frank A. Zagara of the Kodak staff. Loaded with a film cartridge, its price is $15.95, and it will be produced until 1966.

07-16. Dictaphone Corporation introduces this *Time-master,* designed by Joseph M. Parriott, Joseph Rinaldi, and Douglas Long of Raymond Loewy/William Snaith, Inc.

07-17. The Hoover Company introduces this *Dial-a-Matic,* the first cleaner combining the functions of both an upright and a canister cleaner, designed by consultants Smith, Scherr and McDermott and staff designer Carroll Gantz.

07-18. Xerox Corporation introduces the first desktop copier that makes copies on plain paper, the 813, designed by James Balmer, Jr., and William Armstrong, of Armstrong/ Balmer & Associates, Inc. See1959.

07-19. Braun introduces the HT 2, an architecturally clean, slim-line toaster with a two-slice tandem arrangement, designed by Reinhold Weiss of Braun staff.

07-20

07-21

07-23

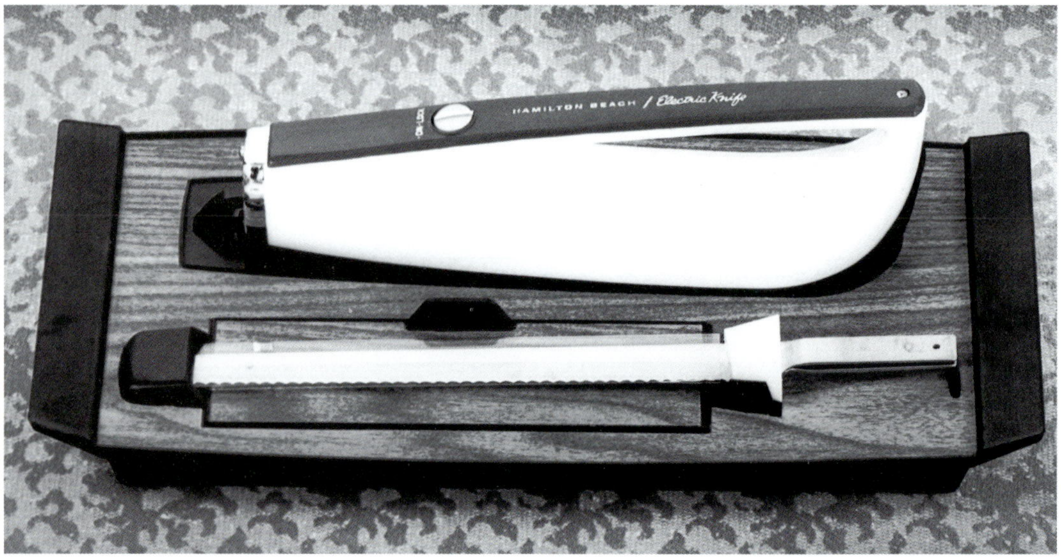

07-24

1964

The New York World's Fair in Flushing Meadows, New York, begins with a design controversy over its huge symbolic *Unisphere*. U.S. Steel, the *Unisphere's* builder, consults with Peter Müller-Munk but actual design is by Gilmore D. Clarke. Walter Dorwin Teague Associates designs a monorail. General Motors' *Futurama*, designed by Henry Dreyfuss Associates, is the most popular exhibit. *Scholar's Walk*, an exhibit designed by Charles Eames, is in the IBM pavilion designed by Eero Saarinen. Critics call the Fair the "design fiasco of the decade."

Kaufmann Awards (see 1960) are revised to become biennial. Awards announced this year are to L. Bruce Archer and Kenneth Bayes of the UK; Jay Doblin and Frank J. Tyson of the U.S.; and GK Industrial Design Association of Japan.

Westinghouse establishes a Consumer Products Design Center (CPDC) in Columbus, Ohio, and appoints Cal Graser as director of industrial design. The CPDC includes the Columbus Group managed by Wallace H. Appel; the Mansfield (Ohio) Products Group managed by Raoul A. Lambert; the Television Group in New Jersey under John Benty; the Home Equipment Group; and the Lamp Group, with Walton E. Sparks managing Advanced Industrial Design. See 1960, 1967

Walter Herbst and Ralph LaZar organize Herbst LaZar Design in Chicago; its first employee is Randy Bell. The firm started in 1962 as Sherman Herbst (Bill Sherman) that merged in 1963 with Podall Design (Podall, Herbst, LaZar Design). In 1978 it will become Herbst, LaZar Bell Design, and by 2000 will be one of the most prominent firms in Chicago.

The Industrial Designers Institute (IDI) presents its annual awards for Ford's *Mustang* (see illustration) for the *Bowlene* container (see illustration), and to Richardson/Smith for Crown Controls pallet truck. Certificates of Design Merit include Xerox's *813* copier (see illustration 1963), and Hoover's *Portable Cleaner* by Smith Scherr & McDermott and Hoover staff designer Carroll Gantz.

07-22

07-20. Climalene introduces this *Bowlene* container designed by Dickens, Inc.

07-21. Stanley Works introduces this *Powerlock* tape rule with chrome housing which will set this product category's typeform for the rest of the century. It is designed by consultant Laird Covey and Stanley's research and development manager Robert West. The shape of the *Powerlock* rule is a registered trademark of the Stanley Works.

07-22. Kodak introduces this low-profile Model 800 *Carousel* slide projector, designed by David E. Hansen of the Kodak staff headed by Arthur Crapsey, Jr. An earlier Model 550, designed by Richard Olson of Kodak's staff, appeared in 1961. The Model 800 will be produced through 1972 and will establish the typeform for slide projectors for the rest of the century.

07-23. The Hamilton Beach Manufacturing Company introduces this 270 T electric knife with a sculptured "hole in the handle," designed by Chapman, Goldsmith & Yamasaki, Inc.

07-24. Ford's new *Mustang* is designed by Joe Oros, head of Ford styling studio, Eugene Bordinat, vice president of styling, L.D. Ash, G.L. Hauserman, C.H. Phaneuf, Damon C. Woods, John Najjar, and J. B. Foster of Ford's design staff. It will become a classic; Jay Doblin will hail it as "one of the great marketing decisions of our time."

07-25

07-26

07-27

1965

The National Endowment for the Arts and Humanities is established by the Federal Aid to the Arts Act signed by President Lyndon Johnson and funded with a three-year appropriation of $63 million. It will evolve into a Design Arts program (see 1967) providing grants for many innovative design proposals, but will be curtailed severely by the 1995 federal budget-cuts.

The Industrial Design Society of America (IDSA) is formed by the merger of IDI (see 1951), ASID (see 1944) and IDEA (see 1957), becoming the single voice of industrial design in the U.S. John Vassos is elected its first chairman of the board, and Henry Dreyfuss its first president. The merger was conceived in 1954 by SID vice president Robert H. Hose and IDI president Robert Gruen, and implemented by Hose and Joseph Parriott of IDI.

IDSA presents annual design awards initiated by the former IDI (see 1951). These will be the last national awards until the Industrial Design Excellence Awards (IDEA) begin in 1980. Awards this year include an overhead projector by R. Williams and W. Sydlowski for Keystone Camera Company; it will set the typeform for such projectors for the rest of the century.

Design USA, a trade exhibit sponsored by IDSA and the Package Designers Council (PDC; see 1952) in collaboration with the U.S. Department of Commerce, displays 300 U.S. products, packages, and corporate identities in London.

Sony introduces consumers to the first portable camera and video recorder. The system used is *Betamax*, which limits recording time to one hour. Large markets will not develop until Philips develops the VCR (videocassette recorder) system.

General Electric's appliance division design staff includes Arthur BecVar as manager of industrial design, major appliances, with a staff of 45 in Louisville, Kentucky; Olle Haggstrom, manager of housewares industrial design, with a staff of 21 in Bridgeport, Connecticut; an electronics staff of 21 headed by Raymond L. Ryder, the Hotpoint staff of 13 headed by Raymond C. Sandin, and other design groups.

Birkenstock sandals designed by Karl Birkenstock are introduced. They are sold only in health food stores. Originally planned for "anatomically correct" post-operative purposes, the shoe will soon become a fashion statement for the youth alternative lifestyle. By 1991 the sandals will be sold through major retailers.

Mobil Oil Company puts into service a cylindrical gas pump designed in 1964 by Eliot Noyes, with graphics by Chermayeff & Geismar.

07-25. The *Trimline* telephone is introduced. Designed by Donald M. Genaro of Henry Dreyfuss Associates in collaboration with Western Electric staff, the phone combines receiver, transmitter and dial into a single form that nests into a compact base.

07-26. Lockheed's SR-71 *Blackbird,* the world's fastest, highest flying plane, designed by Clarence "Kelly" Johnson of Lockheed's "Skunk Works," enters service. Speed is over 2000 mph, and it can reach an altitude of eighty-five thousand feet. It is the successor to the ill-fated U-2.

07-27. The Samsonite Corporation introduces this new *Royal Traveller* attaché case designed by Clair Samhammer of Design West, Inc. See 1962.

07-29

07-30

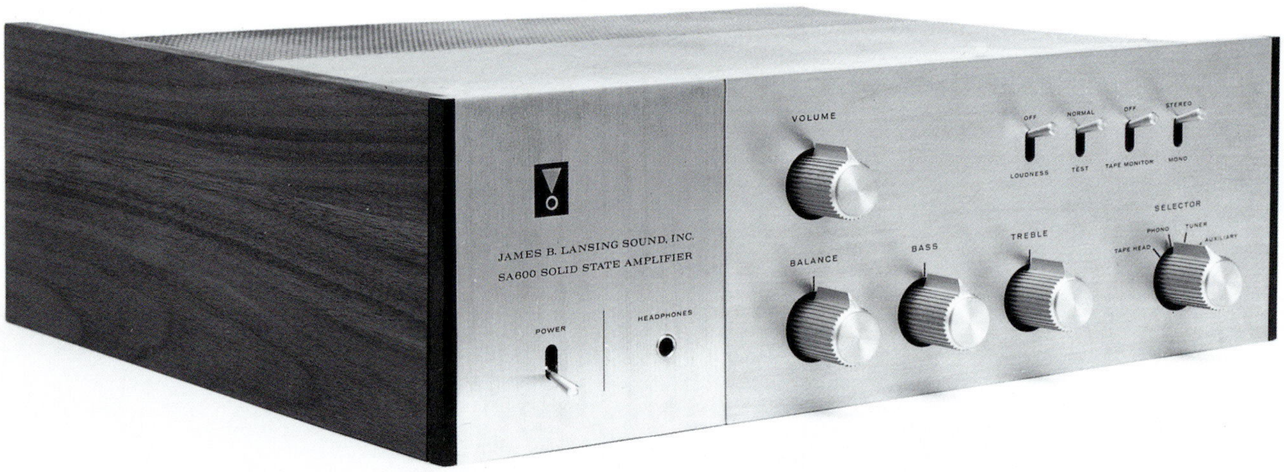

07-31

07-32

1966

The National Housewares Manufacturers Association (NHMA) initiates and sponsors *Design in Housewares* awards. Judges include Gene Bordinat, Jon W. Hauser, Tucker P. Madawick, Arthur J. Pulos, and Harold Zierhut of the National Design Center Board of Design. Ninety-four awards include a Hamilton Beach electric knife (See illustration 1964), a Braun toaster (see illustration 1963), and a cordless electric toothbrush by Mel Boldt for National Presto Industries, Inc. NHMA will continue the awards in 1968 and 1970.

In the USSR, the U.S. Information Agency (USIA) holds an exhibition, *Hand Tools USA,* designed by George Nelson. See 1967.

Armco Steel initiates an annual student design award program that will continue through 1982, involving 900 students from 32 schools.

Joseph M. Parriott, former manager of product design for Raymond Loewy/William Snaith, is named as chairman of Pratt Institute Art School, replacing Rowena Reed Kostellow.

Eudorah Moore, the new curator of design at the Pasadena Art Museum, modifies the format of the *California Design 8* exhibition to include selected one-of-a-kind crafts. Juried industrial design submissions also include prototypes and production models, with emphasis on products for domestic and individual use. The *California Design* series began in 1955 focused on crafts and home furnishings. The new format will continue until the museum becomes the more traditional Norton Simon Museum in 1976.

Philips Electronics, the large Dutch electrical product manufacturer, appoints Knut Yran as head of its "Concern-bureau Vormgeving" established in 1960. Yran will produce a house style design manual in 1973, and will rename his staff as the Concern Industrial Design Centre (CIDC). See 1980, 1989.

General Electric introduces an "Avocado" color on its major appliances. This will soon release a flood of Avocado small appliances and housewares by other makers. Next year, GE will introduce "Harvest," which will have a similar impact on the market. Typically, the new colors on major appliances are airbrushed with a darker tone along the edges to suggest an antique finish, like the most popular version, "Coppertone." See 1955, 1977, 1986.

Presentations begun last year by F. Eugene Smith titled "Why Ugliness, Why Not," rally the design community to encourage civic involvement that will improve the built environment. The visual chaos caused by billboard signage and utility services is a particular concern.

07-33

07-29. Volkswagen introduces its "Squareback" models 411 and 412.

07-30. The Western Electronic Show and Convention (WESCON) confers an Award of Merit to this JBL Solid State Amplifier, designed by consulting designer Arnold Wolf (b. 1927) for James B. Lansing Sound, Inc. Within four years, Wolf will become president of this international manufacturer of high-fidelity systems.

07-31. Sunbeam Corporation launches a "Style Line" of five percolator coffeepots designed by DeFano-Kallman Design, Inc.

07-32. This new *Exxon* logo by Raymond Loewy/William Snaith is implemented, as is a new Celanese logo by Saul Bass & Associates, Rohm and Haas logo by Lester Beall, and Eastern Airlines mark by Lippincott & Margulies.

07-33. Waring Products, Inc. (see 1937) launches this food blender with new solid-state electronic controls, designed by Monte Levin.

07-35

07-34

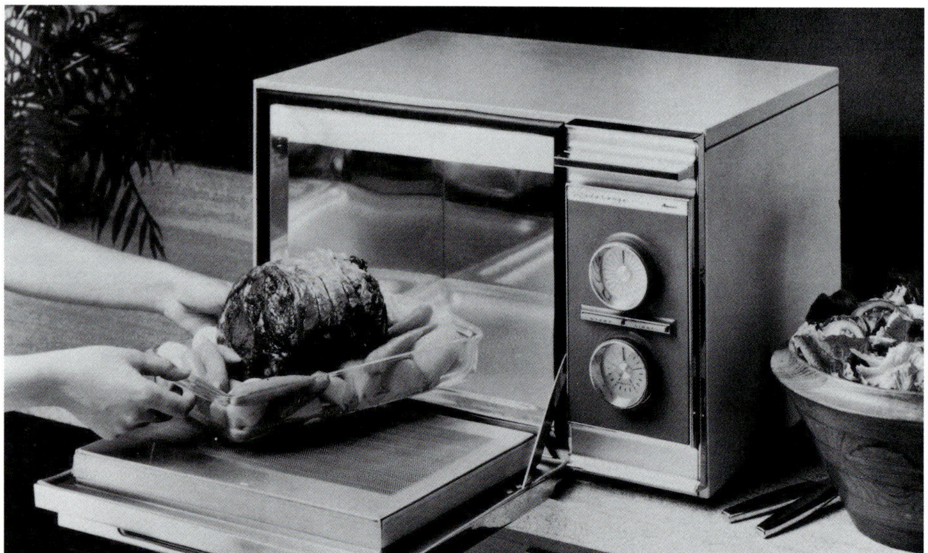

07-36

1967

The fourth *Industrial Design USA* exhibition, designed by George Nelson, is sent to the USSR, sponsored by the USIA. See 1959, 1966. Similar exhibitions will continue until the demise of the Soviet Union in 1991.

The Exposition Universelle (*Expo 67)* is held in Montreal, Canada. The theme is "Terres des Hommes" (Man and his World). Main attractions include a 250-foot diameter, 20-story high geodesic dome by R. Buckminster Fuller housing the U.S. pavilion, for which he will receive the World Medal of Architecture next year.

Westinghouse establishes a Corporate Design Center (CDC) in Pittsburgh, Pennsylvania, with annual funding of one million dollars, upon recommendation of Eliot Noyes. E. W. "Pete" Seay is appointed director, and R.E. Huppertz is made manager of design coordination. Within a year, design staff will grow to 45, but the integration of this new Center with the existing Consumer Product Design Center (CPDC) in Columbus and Mansfield (see 1964) will prove difficult. See 1974.

Owen Finlay Maclaren (d. 1978), a retired aeronautical engineer, revolutionizes baby stroller design with his lightweight aluminum, small wheels, up-turned handles, and simple X-shaped hinges that allow it to be folded into a flat, narrow package. Patented in 1965 and built by his company, Maclaren Ltd., the design will be a huge commercial success.

The National Endowment for the Arts (see 1965) establishes a Design Arts Program to fund grants in art, architecture, and design.

Plastic containers for milk and other foods become more commonplace, as the Food and Drug Administration approves a tin-based stabilizer for polyvinylchloride plastics. This chemical process results in clear plastic containers that are preferable for soft drinks and other food items.

An inflated clear plastic polyvinylchloride "Blow" armchair designed by Gionatan De Pas (1932-1991), Donato D'Urbino (b. 1935), and Paolo Lomazzi (b. 1936) is debuted by Zanotta, Nova Milanese, Italy. It is the first practical furniture supported by air pressure, made possible by a method of electronic welding by a radio frequency to seal the seams.

07-34. Verner Panton designed this stacking chair in 1960, the first single-piece plastic chair and made of fiberglass-reinforced polyester plastic, but technical production problems delayed the chair's production until this year. The chair is manufactured by Vitra GmbH of Basel Switzerland for Herman Miller International and will be produced until 1975.

07-35. These stainless steel scissors with ergonomic handles, designed in 1960 by Oluf Bäckström (b. 1922) and the engineering team of Rolf Lund and Olavi Lindén, are introduced in the U.S. by Oy Fiskars A/B, Finland. The design is similar to an earlier one, highly-publicized in 1952, by Czechoslovakian industrial designer Zdenek Kovar. The orange color occurred because the molding machine that made the prototype just happened to have orange plastic in it, which pleased the Fiskars office staff. Later, a version with red handles will be produced for left-handed people.

07-36. Amana Refrigeration, a subsidiary of Raytheon Manufacturing Company, introduces the first commercially successful compact microwave oven, the *RadaRange,* a 115-volt countertop model retailing for $495. It can cook hamburgers in 35 seconds.

07-37

RCA

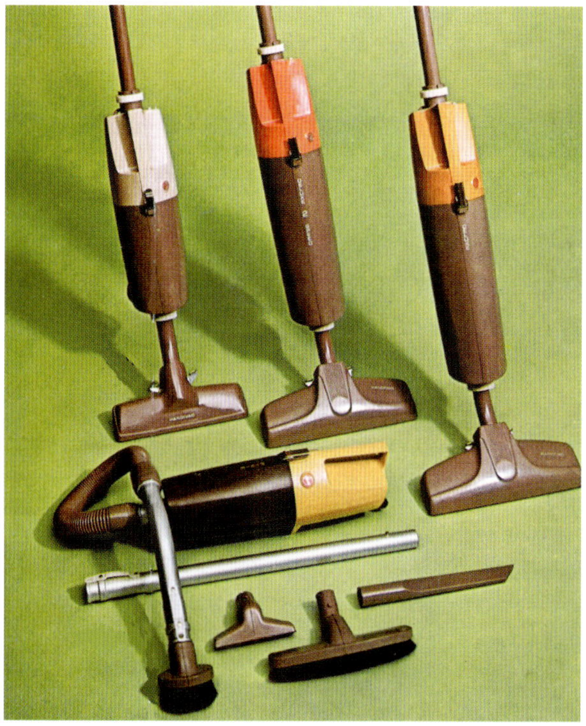

07-38

07-39

07-40

1968

Douglas Engelbart (b. 1925) demonstrates the computer mouse, a device that controls the cursor on a computer monitor, and *Windows*, a display system based on a concept developed in the early 1960s by Ivan Sutherland, an MIT graduate student.

Magic Markers are introduced to the art supplies market. They will soon replace pastels as the primary tool for industrial designers' presentation drawings.

NHMA *Design in Housewares* awards (see 1966) include the *Water Pik* Oral Hygiene Appliance, developed in 1966 by Dr. Gerald M. Moyer, and designed by Emerson, Johnson Mackay for Aqua Tec Corporation; and elegant melamine plastic Hutzler *Cook 'n Serve* tools which will still be popular in 1995.

Tucker Madawick is appointed by RCA as vice president of the Advance Design Center in Indianapolis, Indiana. Madawick leads a staff of 50 at the Center, which was organized upon relocation of the design operation here in 1961.

ALCOA Student Design Awards include a wheelchair that assists the user to rise to a standing position, designed by Peter Bressler of the Rhode Island School of Design.

The *XIV Triennale* in Milan includes a prototype of the "Additional" seating system designed by Joe Columbo. The system is comprised of standardized jersey-covered slabs of polyurethane foam of different heights that can be pegged together on metal frames to create an infinite number and sizes of furniture. The system will be put into production next year.

The Rat für Formgebung, a German industrial design organization, establishes "Gute Form" Bundespreis (national prizes) for industrial design.

Corning Glass Works introduces a "Counter-That-Cooks" electric range, made with a *Pyroceram* cooking surface, designed by Corning staff designer Jerry E. Wright. Four heating elements are hidden under the smooth glass surface. Corning provides a set of *Cookmates,* specially designed cookware with ground and polished bottoms that assure perfect contact with the heat source. By the 1990s, all major manufacturers will have a similar black "glass-top" range.

Suzanne Vanderbilt is promoted to senior designer at General Motors and named the "top woman stylist" in the industry. She will serve until 1977, applying ergonomics to interiors.

07-37. Robert Sarnoff announces "a complete and systematic change in style" for RCA as it launches a new corporate identity program, including this logotype developed by Lippincott & Margulies (see 1945) and Ford & Earl Design Associates for new offices and plants.

07-38. These stick-type (Model 2900) and hand-held (Model 2800) vacuum cleaners designed by Robert H. Hose, of Robert H. Hose Associates, and Carroll Gantz, of Hoover staff, are introduced by the Hoover Company.

07-39. The first moderately priced American sports car from Detroit, the AMX, is introduced by American Motors. It was designed by C. Mashigan and Robert Nixon.

07-40. Herman Miller's *Action Office II,* a refined version of *Action Office I* (1964), is redesigned under the direction of Robert Blaich and introduced. The concept will greatly influence office interiors and be widely imitated. After 1972 it will be redesignated as *Action Office*, and after 1986 as *Action Office Encore.*

07-41

07-42

07-44

1969

IDSA's conference in Washington D.C., "Beyond the Object," is chaired by John Christian. Bill Hannon presents the first multimedia show on the emerging role of the computer in design, which is funded by an IDSA Walter Dorwin Teague Research Grant.

Design in Steel awards (see 1963) this year include a transmission tower designed by Niels Diffrient and D. St. Laurent of Henry Dreyfuss Associates for Southern California Edison Company; and a subway transit vehicle designed by Sundberg-Ferar, and built by the E. G. Budd Manufacturing Company for the New York Metro Transportation Authority.

The Advanced Research Project Agency (ARPA) Net, a network of University computers, is born. The ARPA was formed in 1957 by President Eisenhower to coordinate research, upon the Soviet Union's launch of *Sputnik*. The director of ARPA's Information Processing Techniques Office, Robert W. Taylor, proposed the creation of the Arpanet, which in 1985 will be renamed the Internet.

Corian™, a stain resistant, scratch-and-burn-resistant, non-porous material for countertops, is introduced.

Sony introduces the first videocassette recorder (VCR) system, the *U-Matic*. CBS Laboratories develops an Electronic Video Recording (EVR) Player for home use, designed by John Wistrand of CBS staff, and produced by Motorola, Inc.

Bang & Olufsen, Denmark, introduces its *Beosystem 1200* high-fidelity sound system. It is designed by Dane Jakob Jensen (b. 1926), who created the Bang & Olufsen look by replacing rotary knobs with buttons and sliding switches.

Kenneth L. Thompson and Dennis M. Ritchie, of Bell Laboratories, invent *Unix*, the first portable computer operating system (usable across a wide range of devices). These basic elements will enable advances in computer hardware, software, and networking systems.

Zanotta, Nova Milanese, Italy, introduces the *Sacco* seat, a large leather bag filled with polystyrene beads, designed by Italians Piero Gatti (b. 1940), Cesare Paolini (b. 1937), and Franco Teodoro (b. 1939). It will lead to scores of "bean-bag" chair imitators.

A standard stovepipe elbow found in hardware stores is transformed into colorful houseware products (clock, bookends, clock, and mirror) designed by R. Fleck and Gene Tepper of Tepper/Steinhilber Associates for Elbo Products, Inc.

07-43

07-41. General Electric introduces this side-by-side refrigerator-freezer, with an ice-cube and chilled water dispenser placed in the outside of one door.

07-42. NASA's *Apollo 11* Lunar Excursion Module (LEM) lands on the moon with astronauts Neil A. Armstrong and Edwin E. "Buzz" Aldrin, Jr., while Michael Collins pilots the command ship. It was launched on its journey by a giant Saturn V rocket designed by Arthur Rudolph, once technical director at the Nordhausen, Germany, site that launched V-2 rockets at London (see 1942).

07-43. This new Bell Telephone System logo and corporate identity, designed by Saul Bass, is first used in advertisements and will be adopted by all 22 telephone companies in the Bell System during the next few years.

07-44. The Boeing 747 "jumbo-jet," 231 feet long, makes its first commercial flight. Walter Dorwin Teague Associates designed the jet's interiors. The plane can carry 312-490 passengers and will go into transatlantic service in early 1970 for Pan American Airlines. In 1988, a later Model 747-400 will seat 420-566 passengers. See 1958.

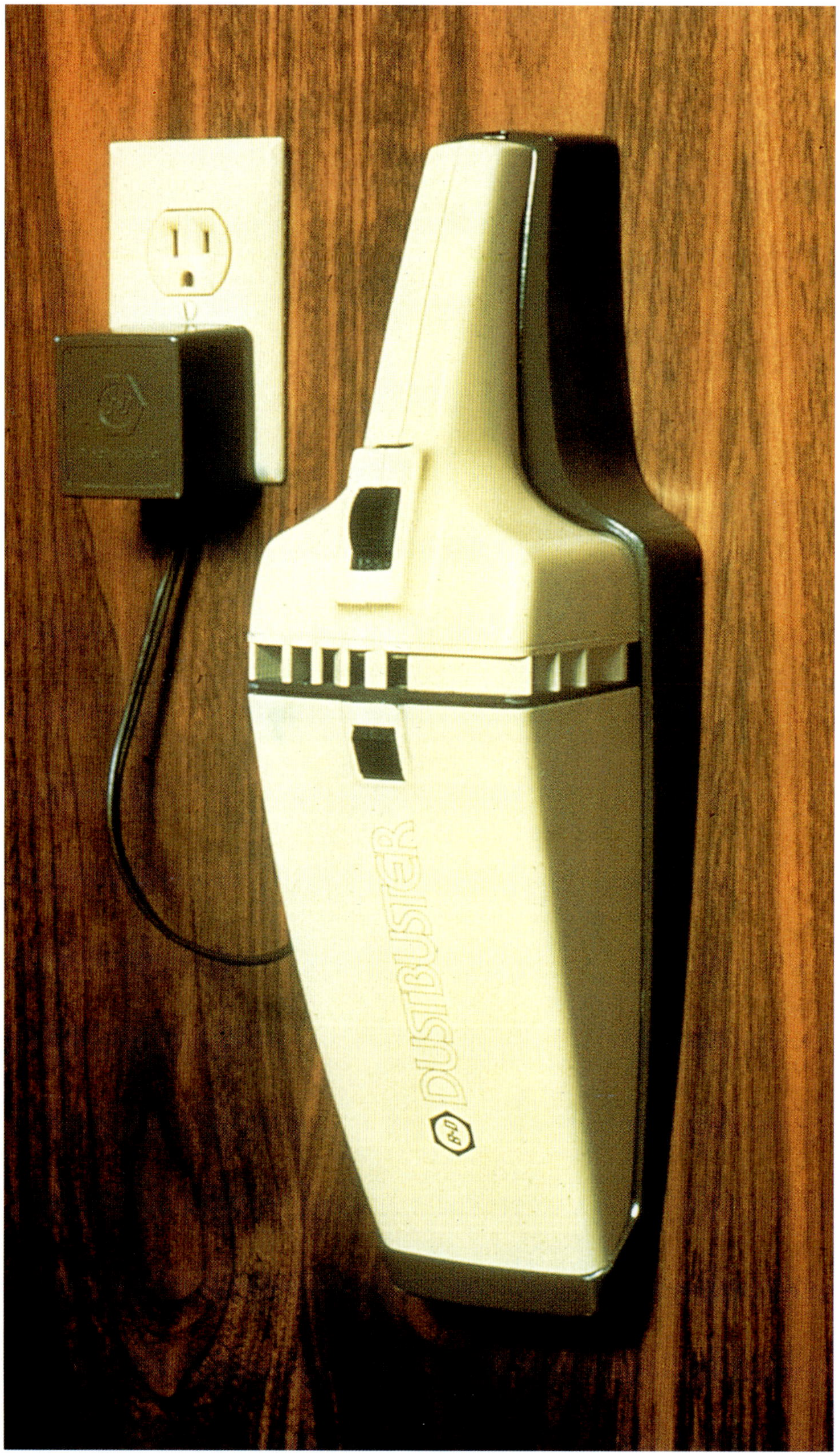
DUSTBUSTER

Chapter 8

1970-1979 Design Diversity

Searching for meaning in the shadow of the Vietnam War, many designers respond to a broad spectrum of social issues: environmental problems, energy conservation, product safety, consumer rights, design for the handicapped, design for the Third World, and design education.

Other designers respond to the changing climate in U.S. business, where many are using design to expand, or at least protect, their shrinking markets against the growing presence of imported goods. Many designers embrace the emerging new technologies resulting from the applications of computers in the space exploration program.

The automotive industry is the hardest hit by foreign competition, particularly from Germany and Japan. Not only are these imported cars more economical and less expensive, they are stylistically more innovative. Automotive design, the traditional trendsetter, clearly has failed in the competitive global marketplace. Detroit's tenacious adherence to the styling clichés of jet fins and rockets is rendered irrelevant after the finless, "non-streamlined," LEM vehicle leaves its tracks on the moon.

With the end of the Vietnam War in 1975, the design community, like the rest of the country, is left in a state of technological, emotional, and generational confusion. One thing is certain—design is changing too.

The general ambiguity of the times brings reforms to design practice. Design offices become less autocratic. Younger designers are given more responsibility and authority in a more youth-conscious culture, and more women enter a field long dominated by males.

Governmental attention to social agendas, environmental issues, and consumer safety leads to design involvement as well. The National Endowment for the Arts fosters an array of design-related projects and initiates the First Federal Design Assembly. A National Design Museum, the Cooper-Hewitt, is established in New York and integrated with the Smithsonian Institution. Celebrations of America's Bicentennial revive a national spirit of patriotism and nostalgia that had been dormant for twenty years. Historic restoration and preservation thrive.

The architectural community rises up against "international design," which in the last 30 years has led to an urban sea of glass boxes and spare concrete slabs. Industrial design, which has for years followed the "International Style" promulgated by the Museum of Modern Art, begins to challenge these same assumptions. New thinking about visual appearance includes the concept that design should express more "playfulness," rather than rigid devotion to recognizable styles.

It is becoming clear that past models of the Bauhaus and international design are obsolete in the cultural momentum of space travel, computers, and social change. The pattern of continued development of new products for transportation and the home, which fueled the success of industrial design since World War II, seems restrictive and mundane. Designers want to solve new problems.

Inventors, innovators, and entrepreneurs are now seen at the leading edge of design. The new heroes are creative spirits like Steven Jobs and Paul MacCready. Tired of continued restyling of everyday household products, designers seek out new territories to explore, such as sports equipment and apparel, products for the disabled, medical equipment, environmental products, musical instruments, and, of course, new electronic and computer products.

08-02

08-03

08-04

1970

The Industrial Designers Society of America (IDSA) initiates its School Recognition Program, aimed at improving the quality of industrial design education by establishing minimum standards, and regular reviews of individual programs.

The International Design Conference in Aspen (IDCA) is entitled "Environment by Design." Keynote speaker is Stuart Udall, former U.S. Secretary of the Interior.

Earth Shoes are named on an impulse by Raymond and Eleanor Jacobs, as they introduce the shoes in New York coincidentally with the first Earth Day. The shoes, designed in 1957 by Anne Kalsø, a Copenhagen yoga instructor, have a "negative heel," with heels lower than the toes. They were intended to be named "Anne Kalsø Minus Heel Shoes." Sales will peak in 1974 and the company will be bankrupt in 1977, but the shoes will continue to be manufactured by the Roots Shoe Company of Canada.

Early computer aided design (CAD) systems are being developed using time-share computers in universities and businesses. One of these is Advanced Remote Graphic Display Scope (ARDS), which uses a "mouse" (see 1968) to draw on adjustable grid systems.

IBM creates the floppy disc to store data used by computers.

Physicist and music lover James T. Russell (b. 1931), working at Battelle Memorial Institute, patents the compact disc. The disc records and replays sounds using digital light impulses that are read by a laser beam.

The *Boby* cart, a lightweight all-plastic tabaret on casters with swing-out drawers and storage compartments, is designed by Cesare "Joe" Colombo for the Italian company Kartell. Designers wonder how they ever worked without one. Popular with architects and designers, Bieffeplast will produce the cabinet into the 1990s.

The *Datapoint 2200*, an early mini-computer with printed circuits, has all the components (processor, display, keyboard, internal memory, and mass data storage) of what will later be called a personal computer (PC). Designed by industrial designer John Frassanito for the Computer Terminal Corporation in Houston, it is introduced at a price of $5000, and has 8000 bytes of internal memory storage.

08-01

08-01. This new logo for American Express is designed by Lippincott & Margulies (L & M). The L & M designers also develop a prototype for an American Motors dealership. The design for the store features a vertical black pylon topped with red, white, and blue "A" symbols.

08-02. *Corelle LivingWare,* a dramatic new break-resistant dinnerware, fabricated of two different glass compositions laminated into a three layer sandwich, is introduced by Corning Glass Works, which promises to replace free any piece which breaks within two years. Its authentic china-like "ping" quality will demolish forever the low-cost plastic competition. By 1972, sales will pass 60 million pieces and, by 1996, two billion!

08-03. Igloo Corporation introduces this *Playmate* plastic picnic cooler with unique pivoting lid, designed by Marlan Polhemus of Goldsmith Yamasaki Specht, Inc. It will have a long life in the marketplace.

08-04. American Motors Corporation's *Gremlin*, designed by AMC design head Richard Teague, is the first true "compact" car on the market, intended to counter the market growth of imported small cars.

08-06

08-08

08-09

1971

Across the street from the White House, the Smithsonian Institution establishes a new museum gallery for American Design at the Renwick House, renamed the Renwick Gallery. The first exhibit, "Design Is...," curated by Lloyd Herman, features contemporary products.

The Industrial Designers Society of America holds its annual conference at Kentucky Dam Village, Kentucky. Chaired by Carroll Gantz, the program features a presentation by Sir Paul Reilly, Director of the British Council of Industrial Design.

Henry Dreyfuss, representing the American National Standards Institute (ANSI), chairs the first meeting of the International Organization of Standards Technical Committee (ISO/TC) in Berlin which sets international design standards for 145 signs and symbols. In 1972, McGraw-Hill will publish the results in a *Symbol Sourcebook* by Dreyfuss.

General Electric opens a new Applied Research and Design Center in Louisville, Kentucky, for major appliances, which includes the industrial design operation and staff of 50 under Arthur BecVar. Richard Montmeat heads the Consumer Products Group and a staff of 37 in Bridgeport, Connecticut.

Texas Instruments (TI) introduces one of the first hand-held calculators, the *Pocketronic*. Weighing 1.5 lbs. and costing $345, it adds, subtracts, multiplies, and divides. TI project manager Jerry Merryman began the design process in 1965. Other calculators on the market this year include Sharp, Busicom, SCM, Bowmar (designed by Gregory Fossella Associates), and Sanyo. By 1975, an avalanche of pocket calculators will force Keuffel & Esser to stop making slide rules, the traditional engineer's friend since Englishman William Oughtred invented the mechanical tool in 1632.

Texans Thomas N. Geist and George C. Ballas invent the *Weed Eater*, a powered lawn-trimming device that is operated by a person in a standing position and uses a nylon fishing line that cuts grass but not people. It will set the tool's typeform for the rest of the century, and eliminate stooping and manual trimming forever.

Rival Manufacturing Company targets "working women" with its Model 3100 *Crock Pot ®*, an energy-saving electric slow cooker available in Avocado or Flame colors. The product is a refinement of a *Bean Pot* produced by the Naxon Utilities Corporation, which Rival acquired last year.

08-05

08-05. Amtrak, the National Railroad Passenger Association, established last year, unveils its new logo and corporate identity, designed by Lippincott & Margulies (see 1945).

08-06. Westinghouse completes an experimental house and kitchen in Coral Gables, Florida, *Electra '71,* designed by the Westinghouse Consumer Products staff. The kitchen features a dishwasher that can be opened on two sides of an island counter, a countertop induction range, and a storage wall that opens from the garage directly into the kitchen to allow convenient unloading of groceries.

08-08. *Max I* stacking melamine dinnerware, originally designed by Massimo Vignelli in 1964 for a small company in Milan, is introduced by Heller Designs, Inc., founded in 1971 in New York by Alan Heller.

08-09. The Tampa Airport Passenger Shuttle System, implemented by Westinghouse industrial design manager Bas Lee and staff beginning in 1965, and built by the Westinghouse Electric Utility Group in East Pittsburgh, Pennsylvania, begins operations on a one-mile track.

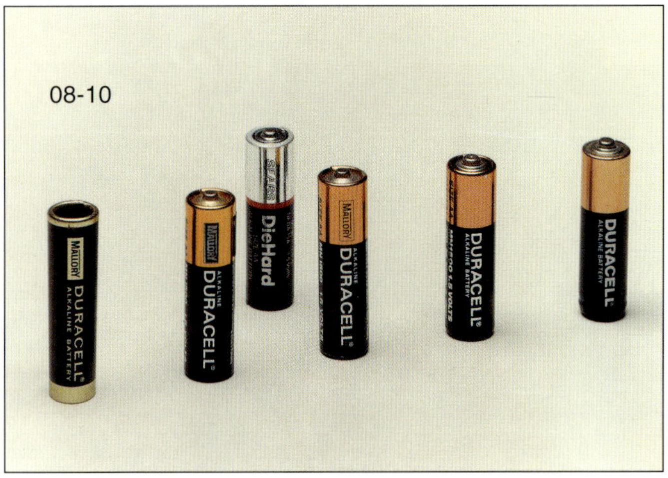

08-10

08-11

08-12

08-14

1972

The Museum of Modern Art mounts the exhibition, *Italy: The New Domestic Landscape*, curated by Emilio Ambasz. Ambasz believes that "we are through with design as object."

The Consumer Product Safety Commission is established by Congress to protect the public from unreasonable risk of injury or death.

Pantheon Books publishes *Design for the Real World* by Victor Papanek. Papanek urges designers to focus on solving third world design problems of survival, and ridicules the extravagant excess of material goods designed by the design establishment. See 1976.

The Central New York Chapter of IDSA holds *Design Dialogue*, a summer conference at the upstate New York campus of Cazenovia College. The conference will become a popular annual design event for many years.

Vincent Marotta and Samuel Glazer of Shaker Heights, Ohio, introduce their *Mr. Coffee* automatic-drip coffeemaker (ADC), which will become ubiquitous in the U.S. thanks to the product's spokesman, baseball legend Joe DiMaggio. Its success will lead in 1995 to *Mrs. Tea,* the first automatic drip hot tea maker, designed by Phil Brookshire. The ADC revolutionizes current coffee brewing methods and becomes the typeform for coffeemakers until the end of the century.

Eastman Kodak Company introduces a *Pocket Instamatic* series of cameras, designed by Kodak staff. It uses a newly developed small 110-size film cartridge. Accessories include a *Pocket Carousel* projector for the small 110-size slides.

Nike Inc. is founded under the name of Blue Ribbon Sports by Portland, Oregon, entrepreneur Philip H. Knight and William J. Bowerman, the University of Oregon's former-track coach. Bowerman designs the first Nike shoes and, by 1990, Nike will be the world's largest maker of sport footwear.

08-13

08-10. Designer Donald Dailey creates a higher quality image for the unique "Coppertop" *Duracell* long-life batteries just introduced by Mallory Battery Company. Dailey was inspired by a trace of rust in his shower water and was reminded that copper means quality electrical conductivity. Dailey also designed the Mallory predecessor "Goldtop" introduced in 1964.

08-11. Braun introduces this KF 20 *Aromatic* coffeemaker designed by Florian Sieffert.

08-12. Teak salt and pepper mills, designed by Jens Quistgaard, are introduced by Dansk. They will become classics, along with many more of his designs, and will continue to be produced into the mid-1990s.

08-13. The *Executive/Memory* pocket calculator, designed and manufactured by Clive and Iain Sinclair of Sinclair Radionics, Ltd., is introduced with a wafer-thin battery and a 7,000-transistor integrated circuit. With slim one-centimeter cross-section, only one-third the thickness of Texas Instrument's *Datamath*, the calculator will win the British Design Council (formerly the Council of Industrial Design) (see 1944) design award in 1973 and will be selected for the Museum of Modern Art's permanent collection in 1974. See 1977.

08-14. Bay Area Rapid Transit (BART), the first new U.S. regional transit system since 1907, goes into service three years behind schedule. Cars for BART, designed in 1965 by Carl Sundberg, R. Figgins and M. Ferar of Sundberg-Ferar, Inc. and Rohr Industries, will set a new typeform for modern mass transit.

08-15

08-16

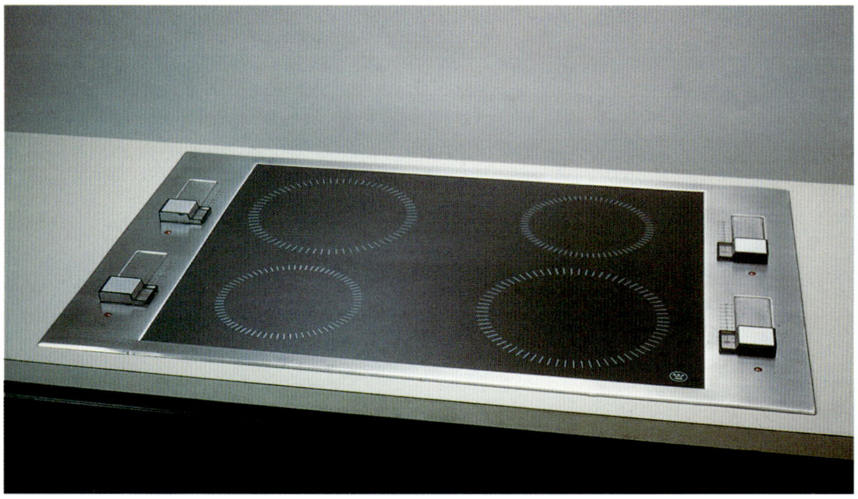

08-17

1973

An international pictograph for handicapped accessibility has been developed by Swedish design students and adopted by Rehabilitation International. Guidelines for application are available from the Committee for Barrier Free Design of the President's Committee for Employment of the Handicapped. The symbol will become universal.

The First Federal Design Assembly is held in Washington, DC, for all federal employees with design responsibility, under the theme: *The Design Necessity*. Nancy Hanks, Chairman of the National Endowment for the Arts (NEA), hosts the event. President Nixon's "Federal Design Improvement Message" states, "There should be no doubt that the federal government has an appropriate role to play in encouraging better design."

Alan Kay, at Xerox's Palo Alto Research center (PARC), develops the first "office computer" based on his *Smalltalk* software, with mouse, icons, and a windows graphic format. See 1974.

IDSA President Arthur J. Pulos announces the January relocation of IDSA's National Office from New York City to McLean, Virginia. IDSA has 723 members and 10 regional chapters.

Under new president James Riddering, Thonet Industries moves to York, Pennsylvania. In 1975, it will launch a new design program directed by Joan M. Burgasser that engages independent consulting designers. In 1979, Gulf & Western Industries will acquire Thonet and its parent, Simmons Company, and, in 1987, Thonet assets will be sold to Shelby Williams Industries, Inc.

An American Iron and Steel Institute design award goes to a wheelchair that allows patients to stand without assistance designed by Peter Bressler, H.A. Wehe, H.W. Wehe, P.W. Gower, G Gustafson, and J.J. Tomlinson of Designs for Medicine, Inc.

Carl Sontheimer (1914-1998), a retired physicist who founded Cuisinarts Corporation in 1971, introduces the *Cuisinart*, a food processor for the home kitchen, at the National Housewares Exposition. Pierre Verdun invented the product in France for commercial use. Verdun modified it for home use and Sontheimer signed a distribution deal with its manufacturer, Robot-Coupe. Because of endorsements by celebrity chefs and gourmet magazines, by 1975 demand will soar and the "Cuisinart" name will become generic for food processors. By 1989 Cuisinarts, Inc., will be sold to Conair Corporation of Stamford, Connecticut. See 1978.

Toyota establishes a design studio, *Calty* (for California Toyota) in California. Other manufacturers will follow; Honda (1975), then Nissan, Mazda, Mitsubishi, Subaru, and Isuzu will form their own California studios as part of a Japanese "Americanization" strategy. See 1986.

08-15. Steelcase, Inc. introduces its *Series 9000* office landscape system designed by Steelcase staff designer Ronald R. Hodges under Jack Hockenberry, who had been appointed director of new products of research and development in a 1970 reorganization.

08-16. This Westinghouse CT-2 Counter Top Induction Range is selected as one of the 20 significant design innovations of the last 20 years by *Industrial Design* magazine.

08-17. The Honda *Civic* is launched in the U.S. after a test export of its N600 model. Designed by Shinya Iwakura and Norimoto Otsuka, the compact front-wheel drive car is welcomed as an economy car. Eight million will be sold by 1996.

08-18

08-19

08-20
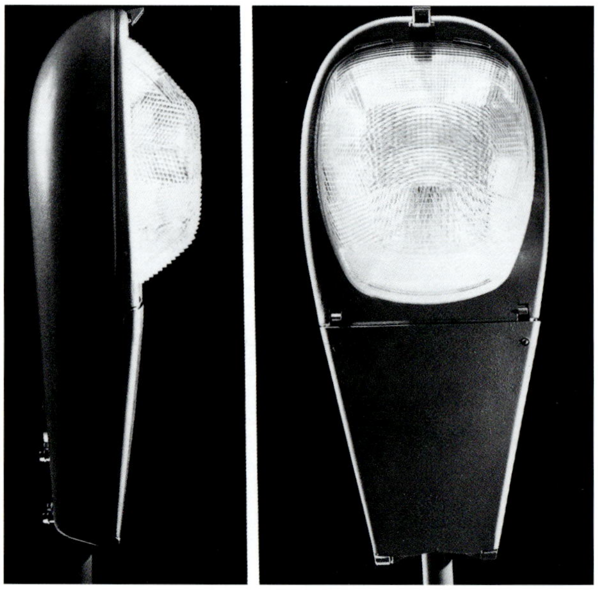

1974

The Design Management Institute (DMI), a forum for corporate design executives, is founded by William J. Hannon Jr. DMI conducts annual conferences on Nantucket Island, Massachusetts. By 1977 it will flourish under the leadership of Peter Lawrence and, after 1985, Earl N. Powell. See 1989.

A prototype of the first personal computer with graphics, the *Altos* desktop, developed at Xerox's Palo Alto Research Center (PARC-), includes overlapping windows, icons, mouse and pointer (WIMPS), and bit-mapped screen as vital elements. Key design participants were Douglas Engelbart (see 1968), the inventor of the mouse, and Alan Kay (see 1973). Xerox will never market the *Altos*, but Steven Jobs will see the prototype during a tour of the PARC.

Westinghouse Corporate Design Center in Pittsburgh is closed, as its Consumer Products Company is sold to White Consolidated Industries (WCI). Current personnel include director J.P. Andrews, and managers L.P. Perfido (architecture); G.W. Smith (graphic design); G.R. Agnew (industrial design), and R.E. Huppertz (planning). In 1975, Wallace H. Appel will become manager of industrial for WCI International, and in 1982, vice president of design, a position he will hold until his retirement in 1991. See 1964, 1967.

Peterson Baby Products and Dave Ellies Industrial Design develop the *Peterson 74* Safety Shell Child Restraint System. The infant seat is designed to meet the proposed automotive standards that will become effective in 1975. Design credit is shared by Ellies designers J. Kain and D. Cone, and C. Farrell, G. Vargo, and M. Swenssen of the Peterson staff.

MIT Press publishes *HumanScale 1/2/3*, ergonomic data compiled and presented by Niels Diffrient, A.R. Tilley, and J. Bardagly of Henry Dreyfuss Associates. See 1982

AB Gustavsberg of Sweden introduces a plastic knife and cutting board, designed by Maria Benktzon and Sven-Eric Juhlin. The designers belong to the Ergonomi Design Gruppen, a research firm funded by governments, institutions, and foundations. The designers developed the product to fully accommodate varying levels of muscle strength and mobility impairments, and to be a form that is visually and physically appealing to the wider public. This approach, termed by the designers as "designing for all," pioneered what will later become known as "universal design."

08-21

08-18. Volkswagen's new *Scirrocco*, *Passat* (known in U.S. as *Dasher*), and *Golf* are all designed by Italdesign's Giorgetti Giugiaro. The *Golf* will become Europe's most popular mass-produced car, and, known as the *Rabbit* (shown in photo), will replace the *Beetle* in the U.S. 14 million will be sold. See 1978.

08-19. Black & Decker introduces its *Workmate* portable workbench. South African industrial designer Ronald Price Hickman designed it and began producing it himself in the UK. A test market last year convinced Black & Decker to pay Hickman significant royalties and to produce it in Canada. More than 55 million will be sold by 2000.

08-20. Westinghouse introduces this OV-15 *Luminaire* street light designed by staff designer Tim Cunningham.

08-21. J.C. Penney's new *Design System,* a comprehensive corporate identity program, launched last year and designed by consultant Jay Doblin of Unimark, and Robert G. Smith and Edward Gorman of Penney staff, wins the IDSA Special Award for the Advancement of Design. See 1982.

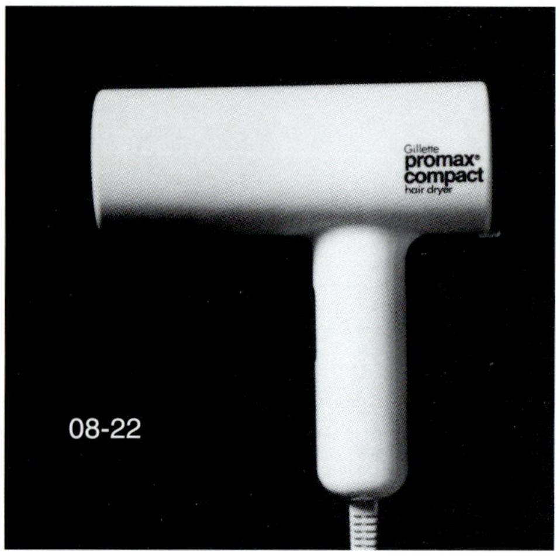

08-24

1975

Microsoft is founded by computer whiz William Henry Gates III, 19, and Paul Gardner Allen, 22. Their first product is a version of the programming language BASIC, invented by John Kemeny and Thomas Kurtz in 1964. Microsoft will make Gates a millionaire before he's 30, and a mega-millionaire by 1995.

The Smithsonian's Renwick Gallery in Washington (see 1971) mounts a retrospective exhibition, *The Designs of Raymond Loewy*, which includes an actual Studebaker *Avanti* (see 1962).

Twenty *Design in Steel* awards include John Deere's 8430 tractors by Henry Dreyfuss Associates, Crown Fork Lift Truck Model 25SC by Dave Tompkins and John Adams of Richardson/Smith, Inc., *Power Loader/Hoe* by Polivka-Logan Designers, Inc. (reformed as Logica Product Development in 1998), and *Comprehensive Seating* for John Stuart, Inc., by William Sklaroff Design Associates.

IDSA honors U.S. Senator Warren Magnuson with a special Government Affairs Citation of Merit for his role in creating the U.S. Consumer Product Safety Act of 1970. Others honored are *Industrial Design* magazine, for its May 1974 article "The Handicapped Majority;" Henry Dreyfuss & Associates for its "Humanscale" project; and IDSA member John Christian for his work on the "Report of the National Commission on Fire Prevention and Control."

The Organization of Black Designers (OBD) is founded in Detroit by David H. Rice to motivate more African-Americans to pursue a career in design. See 1993.

Altair markets a MITS 8800 personal computer kit developed by Ed Roberts. It is sold to electronics hobbyists for $375, but it requires software, additional chips, and a monitor to be fully operational. Most large computer companies do not believe there is any future in such "toy computers."

John Moody of Wisconsin makes the first successful flight from ground level by a powered hang glider. For his mile-and-a-half flight, he used a ten-horsepower engine and propeller, with his legs as a landing gear. The event will initiate a new wave of ultra-light aircraft.

The power of the auto stylist wanes as engineering, governmental regulations, and oil shortages impact car design. Congress passes the Energy Policy and Conservation Act, requiring that cars achieve a corporate average fuel economy (CAFÉ) of 27.5 miles per gallon by 1985, Congress mandates damage-free five-mile-per-hour crashes, and the Clean Air Act is initiated to slow auto emissions, forcing the use of catalytic converters.

08-22. The Gillette Company introduces this *Promax* compact hand-held hair dryer, designed by Michael A. and Morison S. Cousins of Morison S. Cousins + Associates. The Museum of Modern Art will select The *Promax* for its permanent collection in 1977.

08-23. Several new Westinghouse rubber-tired transport systems called "People Movers" enter service at the Miami International Airport and Busch Gardens. They run in linked pairs and were designed by a team headed by Westinghouse lead designer Tim Cunningham. See 1971.

08-24. The first cooperative U.S.-Soviet space mission, the *Apollo-Soyuz* Test Project, is launched, linking a three-man Apollo spacecraft with the two-man *Soyuz 19*.

08-25

08-26

08-27

1976

Cooper-Hewitt Museum becomes the Smithsonian Institution's "National Design Museum" in New York under the direction of Lisa Taylor. The Cooper-Hewitt, founded in 1897 by Sarah, Eleanor, and Amy Hewitt, became a branch of the Smithsonian in 1967, located in the landmark 1901 Andrew Carnegie mansion. Product design collections include archives donated by the Henry Dreyfuss estate.

Design Canada, a National Office of Design in the Department of Industry, Trade & Commerce, is established as the promotional and executive arm of the National Design Council (see 1961) to improve the quality of design in Canada. The Council has produced design case histories, initiated a lecture series in 1973, and has established industry grants for innovation of new products. *Design Canada* will remain the nation's champion of design until about 1983. See 1994.

Designed for the Common Good, an exhibition to promote design awareness, opens at the Design Centre in Brussels in December. The exhibition is commissioned by Belgium's Ministry of Culture through ICSID, and designed by Fulton + Partners, Inc. of the U.S.

IDSA celebrates the American Revolution Bicentennial with its annual conference in Philadelphia, *Evolution/Revolution*, chaired by Peter Bressler. Victor Papanek is "disinvited" as a speaker, because of his criticism of IDSA in his book, *Design for the Real World* (see 1972), but in 1999 will receive IDSA's Personal Recognition Award.

Design Michigan is established by the Michigan Council for the Arts to educate the public about the value and effective use of design. It will function actively until 1991, when the Michigan state government withdraws funding.

Steven G. Wosniac and Steven Jobs raise $1500 and spend six months designing and building *Apple I* computers in a California garage. They will sell 200 of them. Jobs and Wosniac "borrowed" parts from Atari, where they both work. Jobs had also toured Xerox's Palo Alto Research Center (PARC) and seen the *Altos* computer that had been developed in 1974. See 1977.

Daniel F. Cudzic, of the Reynolds Metals Company, patents a pop-top tab for aluminum cans which stays attached, thus solving an enormous environmental problem caused by the original detachable tabs (see 1966).

The *Reach* toothbrush, with an "angled" handle, is designed by Percy Hill and John Kreifeldt, who are both professors at Tufts University and work together at Applied Ergonomics Corporation. The toothbrush is developed by DuPont and introduced by Johnson & Johnson. Squibb introduced a similar "angled" handle toothbrush in 1937, patented by a dentist, Dr. A. Fuller, which used the same angle as the standard dental mirror.

08-25. The *Ergon* chair, with an innovative gas-cylinder post support that absorbs the impact of "heavy" sitters, is designed by William Stumpf for Herman Miller, Inc.

08-26. The Honda *Accord* is introduced in the U.S. By 1993 it will be the best-selling car in the U.S.

08-27. The *Concorde* begins regularly scheduled supersonic service for Air France between Paris and Rio de Janeiro, and for British Airways between London and Bahrain. Both begin service to Washington, D.C. Raymond Loewy/William Snaith office designs the interiors.

08-28

08-29

08-30

1977

The first ICSID-Philips worldwide award of $10,000 for design achievement goes to The National Institute of Design, Ahmedabad, India, for a collection of projects concerned with cooking appliances, birth control devices, and a wheelchair made from readily available parts.

IDSA holds its annual conference at Lake Barkley, Kentucky, chaired by Ralph LaZar with the theme "Patterns: The Sense of Design." Guest speakers include Neil Armstrong, the first man to walk on the moon, futurist Robert Theobald, and commentator Sydney Harris. IDSA initiates the development of Industrial Design Professional Qualifications standards.

The Philadelphia Art Alliance, in collaboration with IDSA, hosts an exhibit, *Industrial Design II...The Art We Live By,* that shows the role of aesthetics in industry.

The New York office of Raymond Loewy/William Snaith closes in Chapter 11 bankruptcy, after 50 years in business. The office is purchased by Sheridan Fundings Corporation, with James Sheridan becoming president of the renamed firm. Loewy still maintains offices in Paris (Compagnie de L'Esthetique Industrielle), London (Raymond Loewy International Inc.), and Sweden.

Terence Conran (b. 1931), who operates a chain of 32 *Habitat* contemporary home-furnishing stores (founded 1964) in Britain, France, and Belgium, opens a similar chain known as *Conran's* in the U.S. By 1981 he will buy *Mothercare*, with 400 stores worldwide, and in 1983 will merge with other chains and become a major enterprise named *Storehouse*.

The plastic soda bottle debuts. It was developed with a new clear plastic, PET (polyethylene-terephthalate), by DuPont chemist Nathaniel Wyeth, the son of famed artist N.C. Wyeth and brother of artist Andrew Wyeth.

George Willig, a young industrial designer frustrated with toy design, designs unique foot clamps, and climbs a tower of the New York World Trade Center. He is arrested and fined a penny for each story of the building he scales.

Steve Jobs (b.1955) and Steve Wosniak (b. 1950) found Apple® Computer Inc. and introduce the *Apple II* ® computer. It is the first mass-produced personal computer, features *Visicalc*, the first spreadsheet program, and will remain the best-selling computer until IBM enters the personal computer market (see 1981).

General Electric eliminates "shaded" colors in its major appliances and introduces its "New Naturals" unshaded palette of Almond, Harvest, Wheat, Coffee, Fresh Avocado, Onyx, and Snow on many of its appliances. The color "Almond" will proliferate throughout the appliance industry and survive the rest of the century as a "neutral" alternative to white. See 1955, 1966, 1986.

08-28. In Britain, the *Gossamer Condor*, a human-powered, bicycle-driven aircraft designed by Paul MacCready and piloted by Bryan L. Allen, wins the Kremer prize of £50,000 by flying a figure-eight around two markers four-fifths of a kilometer apart. See 1979,1987.

08-29. The *Sovereign*, a pocket calculator designed by Iain Sinclair of Sinclair Radionics, is introduced. See 1972.

08-30. Braun introduces the *ET 33* pocket calculator, designed by Dieter Rams and Dietrich Lubs. The calculator is cited by some as the definitive answer to what a calculator should look like and in a size that is not so small as to be inefficient or difficult to use.

08-31

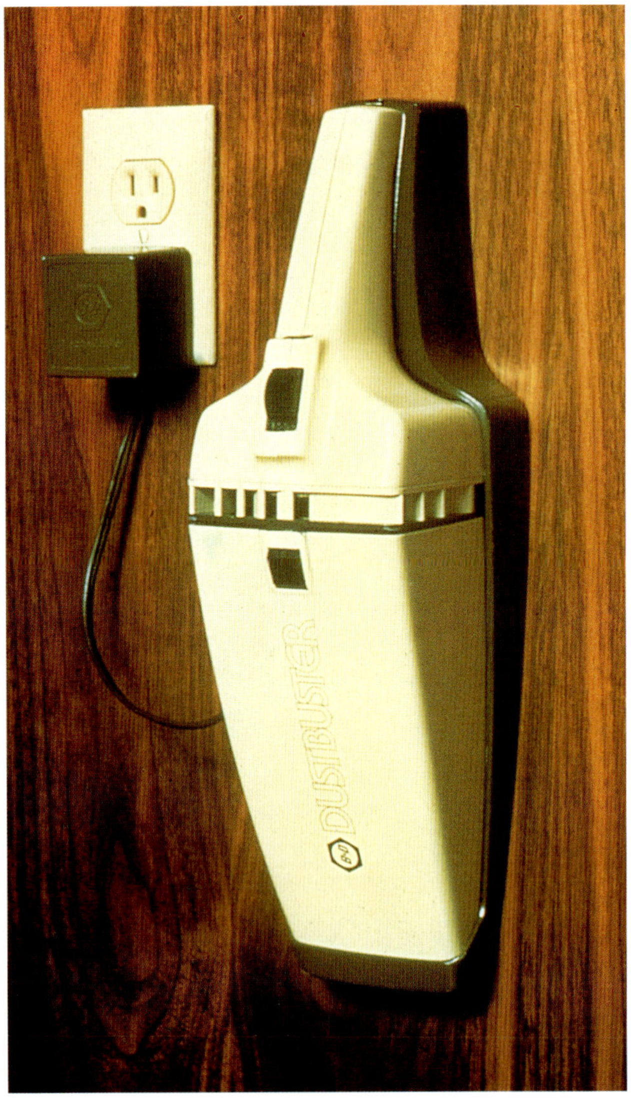

08-32

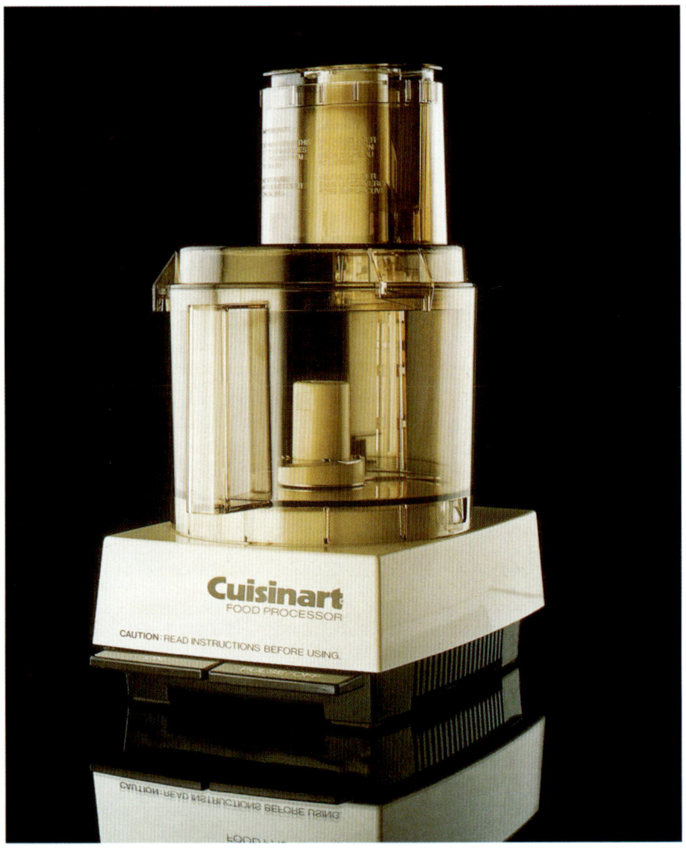

08-33

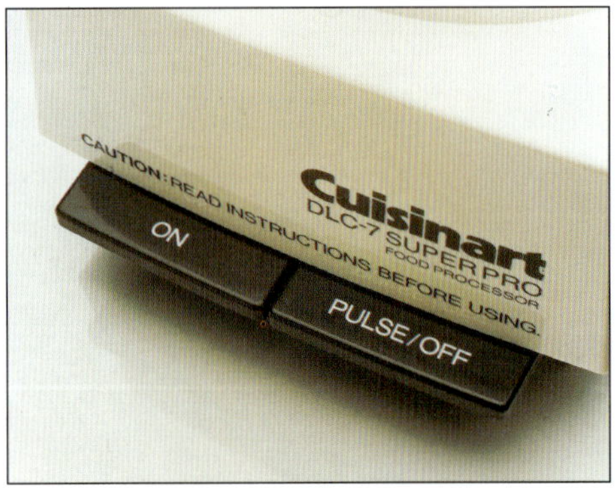

08-34

1978

At IDSA's annual conference, *Thrival: Beyond Survival*, held at Asilomar, California, keynote speaker Ralph Nader urges designers to stand up for safety within their corporations. Awards for corporate product and design excellence are presented to the Danish firm of Bang & Olufsen and to IBM. IDSA's Personal Recognition Award is given to Raymond Loewy for his contributions to the profession.

IDSA adopts a new *Code of Ethics for Industrial Designers,* developed by a task force chaired by Arden Farey, *Entry Level Professional Qualifications for Industrial Designers,* developed by a committee chaired by Larry Bell, and a formal *Definition of Industrial Design*. All will still be valid IDSA standards at the end of the century.

IDSA opens new national headquarters at 1717 "N" Street in Washington DC. With its first full-time Director, Brian Wynne, and staff, Celia Weinstein, IDSA hopes to gain greater public exposure by its proximity to government and other professional and business associations. IDSA has 976 members.

A group of private investors led by James Fulton form a corporation, Design Publications, Inc., to lease the publishing rights to *Industrial Design* magazine from Billboard Publications, which had just acquired Whitney Publications (see 1954).

The Japanese Graphic Designers Association is formed. Kazuo Kashimoto designs its symbol, "JAGDA."

Davin Stowell Associates is founded in New York by Davin Stowell, who will be joined by Tucker Viemeister, Tom Dair, Tamara Thomsen, David Hales, and Dan Formosa, and become Smart Design, Inc.

The first electronic typewriters with memory, Olivetti's *ET* range, designed by Mario Bellini, and one by Casio of Japan, are introduced simultaneously.

98% of US households have television sets, 78% in color.

The last *Beetle* (see 1938) is produced in Germany, as production of the *Rabbit* (see 1974) begins. Since their introduction in the U.S. (see 1949), over 19.2 million *Beetles* have been produced. They will continue to be made, however, in Mexico, Brazil, Nigeria, and South Africa.

William P. Lear designs a new corporate turboprop aircraft structured almost entirely of graphite epoxy, the *Learjet*.

Transportation Mazda Motor Corporation, Hiroshima, Japan, introduces its first-generation RX-7, designed in Japan, and opens a design office in California, Mazda Research & Development of North America, with a single designer from Japan. By 1982, the staff will increase to five and, by 1991, to 28. Most will be locally employed.

08-31. Black & Decker introduces the first cordless hand-held vacuum cleaner, the *Dustbuster®*, designed by Carroll Gantz. It will establish a new product category and typeform, transforming Black & Decker from a power tool maker to a total housewares manufacturer after its acquisition of General Electric's Housewares division in 1984. By 2004, more than 150 million *Dustbusters* will be sold.

08-32 & 33. Cuisinarts, Inc. introduces this *Cuisinart DLC-7* food processor designed by Marc Harrison, which includes many improvements over its predecessor (see 1973), such as easy-to-use push controls, early examples of what will later be called "universal design" (see 1991).

08-34. J.C. Penney introduces this coffeemaker designed by design manager Cooper C. Woodring.

08-35

08-36

08-37

1979

Overlook Press publishes *Industrial Design,* by Raymond Loewy, Van Nostrand Reinhold publishes *In Good Shape*, by Stephen Bayley, and Temple University Press publishes *Twentieth Century Limited: Industrial Design in America 1925-1939*, by Jeffrey L. Meikle.

IDSA's annual conference in Washington, D.C., chaired by Cooper Woodring, is titled, *The Gross National Product.* Featured speakers include Senator Nancy Landon Kassebaum, television's *60-Minutes* commentator Andy Rooney, and Dr. Jordan Baruch, Assistant Secretary of Commerce for Science and Technology. John Vassos is recognized for his outstanding contributions to the industrial design profession.

A one-hour film on the life and work of Raymond Loewy, titled, "Looking Back to the Future," is completed for Public Broadcast Syndication, and includes exclusive interviews with Loewy in France. The film was produced by the IDSA Walter Dorwin Teague Research Trust, enabled by a grant from the National Endowment of the Arts.

Bill Moggridge and Mike Nuttall, both originally from the UK, found the design firm, ID Two, in San Francisco. They will part in 1983, with Nuttall starting Matrix Design.

A stainless steel and plastic espresso coffee maker designed by Richard Sapper is introduced by Alessi of Italy, and is awarded a *Compasso d'Oro*. The two main parts, the top coffee container and the lower water boiler, are easily opened with a handle-operated lever device.

Nike introduces its *Tailwind* running shoe. Nike, founded by Phil Knight, premiers a "swoosh" logo designed by art student Carolyn Davidson. See 1986.

The *Gossamer Albatross,* a human-powered aircraft designed by Paul MacCready, and piloted and powered by Bryan L. Allen, crosses the English Channel (22.5 miles) in 169 minutes, duplicating Louis Blériot's flight of 1909, and winning a Kremer prize of £100,000. See 1977.

Japan is now producing more cars than the U.S., whose U.S. market share is down to 75 percent. It is Detroit's worst year since 1931. Chrysler loses $1.7 billion, Ford loses $1.5 billion, and General Motors $763 million. The new Isuzu *Impulse*, designed by Giorgetto Giugiaro is the first car to employ an important aerodynamic principle by having windows flush with the body form.

08-35. Sony introduces its *Walkman* pocket stereo cassette player at $200. Designed by a team headed by Mitsuro Ida, it was preceded by a small portable tape recorder, the *Pressman,* in 1978. Sony co-founder Masaru Ibuka suggested combining it with lightweight headphones, then under development. When first introduced, Sony thought the name *Walkman* sounded "funny" in English, and initially used the name *Soundabout* in the U.S., and *Stowaway* in UK.

08-36. The *Diffrient Seating Series*, designed by Niels Diffrient, Jeffrey Osborne, Tom Latone, and Hy Zelkowitz, are introduced by Knoll International, Inc. The series includes articulated High and Low Back Executive leather armchairs; Advanced Management Swivel and Tilt chairs with and without arms, and Advanced Operational Task chairs.

08-37. Oneida Ltd., founded in 1848, of Sherrill, New York, introduces Heirloom stainless steel flatware, *Omni,* designed in 1975 by Ellen B. Manderfield. The pattern will be produced until 1991.

STEINBERGER

Chapter 9

1980-1989 Expressive Design

The global design style ideal, which embodies international design quality and enables manufacturing companies to satisfy contemporary market trends, continues. Designers develop a flexible, yet consistent, visual language for broad contemporary product lines that organizes the international personalities of brand names and simplifies the complex, varying national regulatory standards.

Museums address design history with respect and acclaim. After decades of attention to product design, The Museum of Modern Art turns its focus to post-modern architecture. The traditional design canon is symbolically sealed with a 1983 exhibition, *Design Since 1945*, which essentially ends the era in 1970, putting to rest "International Style" product icons with dignity. The *Machine Age in America* exhibition at the Brooklyn Museum likewise retires the 1930s.

The expression of emotion through form and color, rather than the cold, neutral Bauhaus logic of geometry, signals a new interpretation of design excellence. The Memphis design group, of Milan, epitomizes this dramatic shift in 1981, with furniture designs that violate all conventional rules of reason and functionality. A new concept of design semantics encourages the inclusion of whimsy and visual expressiveness into product design. These concepts, accepted by global product manufacturers, enter the mainstream.

The new understanding of design presents designers as businessmen, entertainers, humanists, and technological innovators. Indeed, the computer revolution sees a vast expansion of design potential, with machines as powerful design tools. Concepts for intelligent products, from robots to remote-controlled devices to serve mankind, free the designer from traditional forms dictated by motors and construction limitations, permitting pure forms that better fit human needs.

So too arise new benchmarks for excellence in industry, as international competition forces lower costs and higher quality, and as business experts identify user service and attention to the customer as the keys to commercial success.

The optimistic times encourage design exploration and experimentation. A new public recognition of design appears. After decades of no national awards by design organizations, IDSA implements a national "Design Excellence" awards program in 1980.

By mid-decade, the automotive industry abandons its Harley Earl traditions and embraces the more evocative and sensuous forms reminiscent of the 1930s. Ford risks its reputation on the introduction of the *Taurus* and wins success, capitalizing on its belief that the public is ready for a new look in cars.

In a sense, designers everywhere feel a new freedom from the shackles of history and the oppression of marketing "studies" that encourage imitation instead of creativity. When the *NeXT* computer debuts as a black cube, few are shocked.

While total consensus evades the design community, vigor and freshness in many new design directions result in the collaboration and cooperation of designers with engineers, marketing, manufacturing, and the consumer. In fact, a new optimism pervades the country.

09-01

09-02

09-03

1980

Alvin Toffler publishes *The Third Wave*. The first was agricultural and the second, industrial. Toffler cites 1955 as the beginning of the third wave, or as some call it, the information age. Toffler outlines the many changes in society driven by individuals seeking more diversity and freedom of choice.

IDSA initiates Annual Design Awards for the first time since the IDI awards in 1965. Called the *Industrial Design Excellence Awards* (IDEA), the program is chaired by IDSA's Ralph LaZar and awards are presented in San Antonio at IDSA's annual conference. The first IDEA awards include: *The Diffrient Chair* (see illustration 1979); *The Pocket Socket Wrench*, by Edward Levy, Michael Ballone, and Edward Meisner of Innovations and Development, for New Britain Tool Division of Litton Industries; The *5120 Computing System* by Tom R. Hardy for IBM; Crown's *Narrow Aisle Reach Truck* (see illustration), and the *IM Modular Signage System* for APCO Graphics, Inc. (see illustration).

Robert Blaich is named Senior Managing Director of Philips Corporate Design at Philips Electronics NV in the Netherlands, replacing Knut Yran (see 1966). Responsible for all product, graphic, packaging, and corporate identity, he heads 250 designers in 21 countries. Blaich will radically change the image of Philips, integrating the disciplines of design, engineering, and marketing, and involving international designers in a "global design" philosophy. See 1989.

Arnold Wasserman is named director of corporate industrial design at National Cash Register Corporation (NCR). He will initiate comprehensive corporate product design standards that consider much more than the usual graphic design corporate identity (see 1984).

Minnesota Mining and Manufacturing Company (3M) introduces yellow *Post-it* notes. In 1974 Arthur Fry (b. 1931), a chemical engineer at 3M who was frustrated by paper bookmarks falling out of his hymnal while he sang in a choir, began development of the pads of self-adhesive paper with Spencer Silver (b. 1941), who created a non-binding adhesive.

RFSU Rehab, of Stockholm, Sweden, produces *Eat/Drink* cutlery designed by Ergonomi Design Gruppen (see 1974), following the goals established by two of its members, Maria Benktzon and Sven Eric Juhlin. Specially designed for those with rheumatoid arthritis, the utensils include a "knork" and a "knoon," that are combinations of typical tools.

09-04

09-01. Ronald W. Cobb and Elyse B. Reeves design this *IM Modular Signage System* for APCO Graphics, Inc. It will be named by IDSA in 1990 as a *Design of the Decade*.

09-02. David Tompkins, Frank Wigus, and David B. Smith of Richardson/Smith, Inc., design this *Narrow Aisle Reach Truck* for Crown Equipment Company. In 1990, it will be named by IDSA as a *Design of the Decade*.

09-03. Herman Miller, Inc introduces the *Burdick Group Office System* furniture, designed by Bruce Burdick, that uses a flexible beam system of "executive workbenches." In its January 4, 1982 issue, *Time* magazine will name the furniture system as one of the five best industrial designs of 1981.

109-04. Hamilton Beach Manufacturing Company introduces this *Popaire 5* corn popper designed by Monte Levin Associates Inc.

09-05

09-06

09-07

1981

In Milan, Ettore Sottsass and Italian industrialist Ernesto Gismondi found *Memphis,* a group of avant-garde designers. *Memphis* creates the design media event of the year with garish and irreverent product designs in bold color and laminates that emphasize surface decoration more than function. The *Casablanca* sideboard, the signature piece by Sottsass, becomes the icon of *Memphis*, of postmodernism and of the Italian avant-garde.

IDSA presents IDEA awards (see 1980) in Los Angeles, including those to: *Burdick Group* furniture (see illustration 1980); the *Edge* Videotape Console by Steinhilber & Deutsch, Inc., for CMX Orrox; and the *Steinberger Bass Guitar* (see illustration). Awards are expanded to include Recognition of Design Achievement, one of which is given to The *Hannah Desk System* (see illustration).

IBM enters the personal computer market with a computer that uses a disc operating system (DOS). Designed by a team headed by Philip Estridge, the new software by Microsoft will capture 75 percent of the market, despite being less user-friendly than the Macintosh graphic-based system.

John Z. De Lorean, a former GM manager who quit in 1975 to form De Lorean Motor Company, introduces the new *De Lorean* DMC-12 sports car. Design is by Giorgetto Giugiaro. At $26,000 each, 3000 will sell, yet the company will become insolvent. De Lorean is arrested on charges of conspiracy to distribute cocaine, but later acquitted. The plant in Ireland will shut down in March 1982, after producing 8,583 cars. By 1985, the car will star as Michael J. Fox's time machine in the hit film, *Back to the Future.* In 2000, De Lorean will file for bankruptcy and lose his $15 million home.

Nissan Design International (N.D.I.) of Japan, established in 1979, names Gerald P. Hirshberg director of design at its new international design center in La Jolla, California. Hirshberg had been the chief designer at GM's Buick and Pontiac studios. By 1991 N.D.I. will employ 50 designers, and, by 1995, Hirshberg will be named president.

Olivetti introduces the *Praxis 35* portable electric typewriter designed by Mario Bellini.

09-05. Ned Steinberger designs this *Steinberger Bass Guitar* for Steinberger Sound Corporation. It will be named by IDSA in 1990 as a *Design of the Decade,* and will be listed as one of the five best industrial designs of 1981 in the January 4, 1982 issue of *Time* magazine.

09-06. Bruce Hannah designs this *Hannah Desk System* for Knoll International. In 1990, it will be named by IDSA as a *Design of the Decade.*

09-07. The F-117 *Nighthawk* "Stealth" fighter plane first flies. Designed by Ben Rich of Lockheed's famous "Skunk Works" in the mid 1970s, the plane's angular body shapes and construction materials evade radar. It will be used for combat in Panama in 1989 and the Gulf War in 1991.

09-08. Ludwig Littmann, of Braun AG staff, designs this *MR 6 Multipractic* hand blender. A compact and unique in-line design, the machine is operated vertically with one hand, and sets a new typeform for blenders.

09-08

09-09

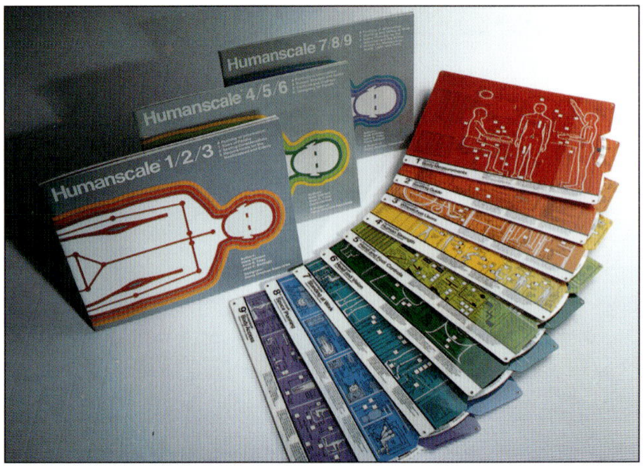

09-10

09-11

09-12

1982

Cooper Woodring is appointed head of design at J.C. Penney, upon the retirement of Robert G. Smith. The department has more than 15 designers. Within three years, J.C. Penney will expand its focus to more profitable fashion apparel, drop its hard lines, and close the design department. See 1974.

Time magazine selects the computer as its "Man of the Year" for 1982, and features a prototypical computer mock-up on its cover, designed and built by industrial design firm Richardson/Smith, Inc.

A compact disc (CD) player, consisting of a digital recording on a 120-mm plastic disc that is read by a laser beam, is introduced in Europe and Japan. Developed jointly by Philips and CBS/Sony, it will reach the U.S. next year, when players will sell for $800 to $1,000, and CDs for $17. The first CD album is *52nd Street*, by Billy Joel.

IDSA presents testimony at a government hearing on a House bill to establish a U.S. Design Council. The U.S. is the only major industrial nation in the world without a federally sponsored design organization. The bill will fail, as will others, including an attempt in 1994.

In Search of Excellence, a book by Tom Peters, is published. Peters writes and lectures on how companies achieve success, not through orthodox business practices, but through creative attention to consumer-related issues, including design.

The Sony FD-2A *Watchman*, a pocket-sized television, is introduced at a price of $280. In the UK, another pocket TV by Sinclair Radionics, Ltd., is introduced at one-third that cost.

The *Itty Bitty* book light is introduced. The tiny portable lamp, designed by Noel Zeller, president of Zelco Industries, which he co-founded with his wife Adele in 1976, will sell nine million by 1994.

C. Josef Lamy of West Germany introduces the *Safari* fountain pen, designed by Wolfgang Fabian of Entwicklungsgruppe, and the *Safari* ballpoint pen, designed by Dr. Manfred Lamy.

The *Sierra* auto, designed by German Uwe Bahnsen (b.1930), vice president of design at Ford Europe, is introduced. The *Sierra* establishes a new direction in design that will strongly influence Ford's *Thunderbird* (see 1983), *Taurus,* and Mercury *Sable* (see 1985).

09-09. *Humanscale 4/5/6* and *Humanscale 7/8/9,* books that present a set of ergonomic design guides by Niels Diffrient, Alvin R. Tilley, and David Hartman of Henry Dreyfuss Associates, are published by MIT Press. See 1974.

09-10. IDSA presents an IDEA design award in Orlando, Florida, to Bill Moggridge (b. 1943) and Steve Hobson of I.D. Two for the design of GRiD Systems *Compass* computer. It will become the first computer selected by the Museum of Modern Art for its permanent collection.

09-11. This *Mighty Mite* vacuum cleaner, designed by Kenneth R. Parker and Samuel E. Hohulin, is introduced by the Eureka Company, founded in 1909.

09-12. Philips Electronics NV of the Netherlands introduces this conceptual "Complete Kitchening" prototype that incorporates tambour-covered storage units for small appliances. The design is by Lou Beeren and Paolo Villa with Robert Blaich as director of design. Philips also sponsors an ICSID award for 1982 and an ICOGRADA award for 1983.

09-13

09-14

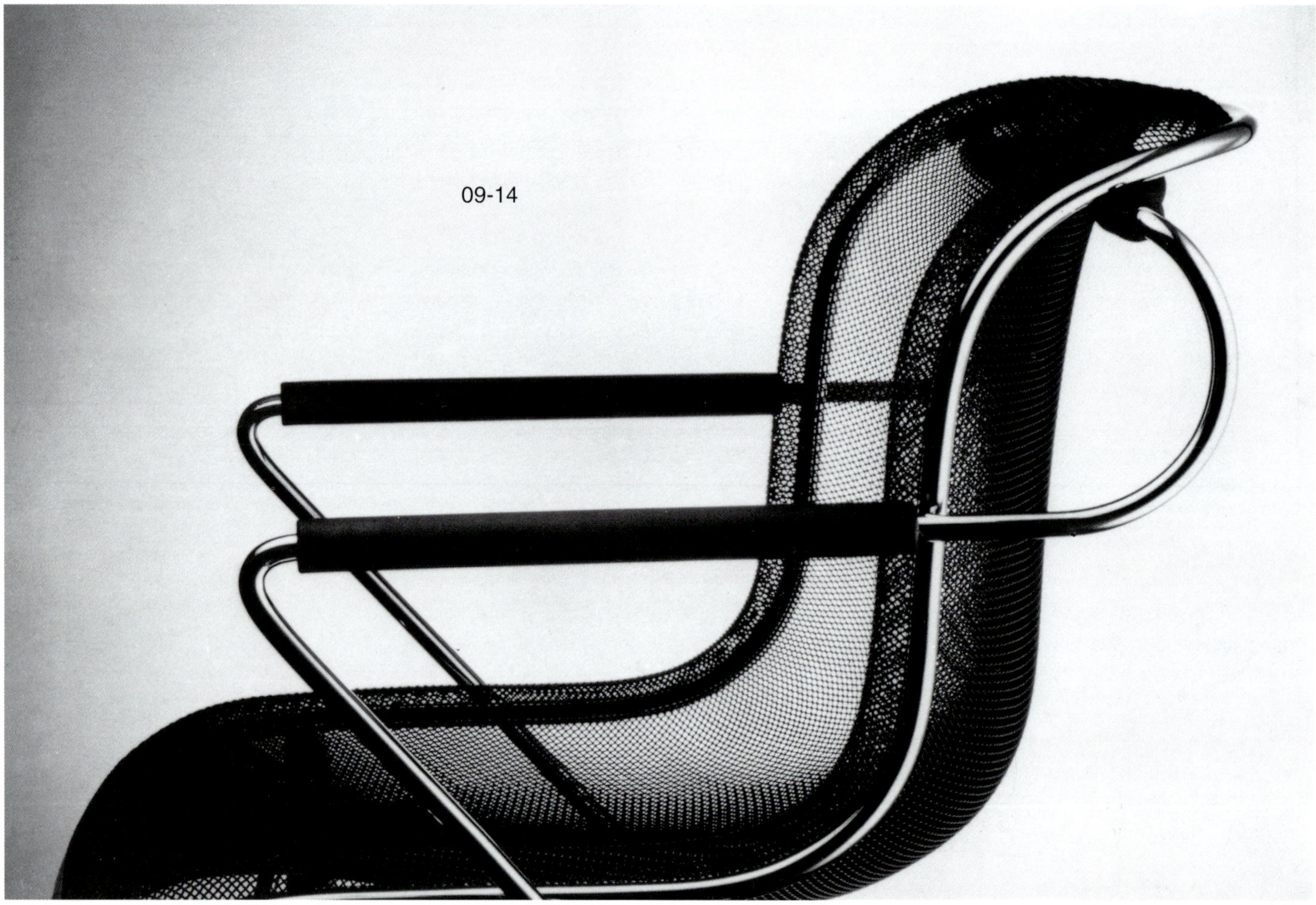

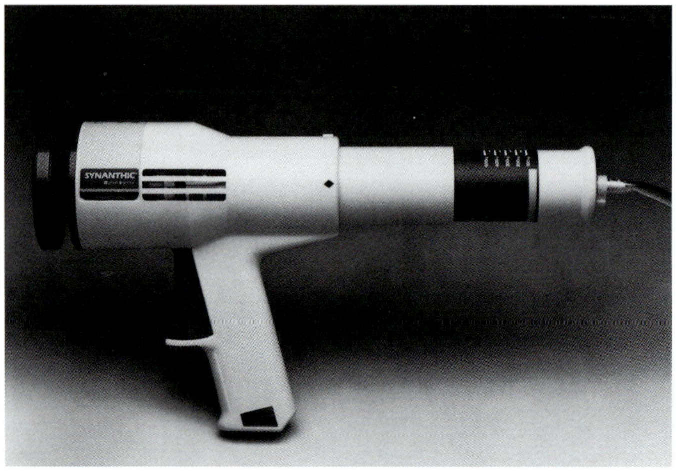

09-15

09-16

1983

The exhibition, *Design Since 1945,* organized by Katherine Hiesinger with installation by George Nelson, opens at the Philadelphia Museum of Art, along with a catalogue of the exhibition by Hiesinger.

IDSA presents IDEA awards in Chicago to: the 1984 Chevrolet *Corvette* by GM staff; the *Penelope* chair (see illustration); the *Compact Diamond Packaging System* by Richardson/Smith Inc. for General Electric; the *IPL 600 Asymmetrical Task Light* by Designspring, Inc. for Ledu Corporation; the *943 Flame Photometer*, and the *Synthetic Rumen Injector* (see illustration).

Katherine J. McCoy, co-chair of Cranbrook Academy of Art, takes office as the first female president of IDSA.

MIT Press publishes *American Design Ethic,* by Arthur Pulos. A second book by Pulos, *American Design Adventure*, will follow in 1988.

Designer Hartmut Esslinger names his design firm in Altensteig, Germany *frogdesign*, a word derived from the initials of the Federal Republic of Germany. He established his firm in 1969, and last year opened a branch office in California in response to Steve Jobs's invitation to work on the *NeXT* workstation (see 1988).

The *Swatch Watch,* designed by engineers Jacques Müller, Ernst Thonke and Elmar Mock in 1980 is introduced at $40. Next year, Max Imgrüth, president of Swatch Watch USA, will introduce innovative colors and patterns that will create a new fashion watch market.

The new Audi *5000,* typifying renewed German interest in aerodynamics, is the first on the U.S. market to present a noticeably "soft" or "rounded" body shape which will be referred to by some as "jellybean" styling.

Ford's new *Thunderbird*, designed by Jack Telnack, North American design section head who had recently served in Ford's European staff, displays what Ford calls the *Aero-look*, which signals a fundamental change in U.S. car styling from the traditional Detroit visual language.

09-17

09-13. Formica introduces *Colorcore*, a solid-color three-dimensional decorative material that is an improvement over previous planar sheeting laminates, at the Chicago NEOCON furniture fair. The idea for such a material emerged from a Design Advisory Board assembled in 1977 by newly appointed design manager Alessandro De Gregori. Designers were invited to enter several design competitions to explore effective uses, and winners include this *Neapolitan* coffee table by Lee Payne.

09-14. The *Penelope* chair, designed by Charles Pollock and Tracey Tooker for Castelli Furniture, Inc.

09-15. Nolan Vogt and Steven Umbach design this *Synanthic Rumen Injector* for Syntex Agribusiness, Inc. The device injects medications directly into the rumen (stomach) of cattle. It will be selected by IDSA in 1990 as a *Design of the Decade*.

09-16. Karl Lagerfeld of Paris introduces "KL" fragrance in this package designed by Marc Rosen Associates of New York. The package will win a "Fifi" award, the fragrance world's annual "Oscar" presented by The Fragrance Foundation for creative achievements.

09-17. The world's first commercial handheld cellular phone, the Motorola DynaTAC™ 8000x, receives approval from the U.S. Federal Communications Commission in 1983. This 28-ounce phone, designed by Motorola's industrial design team led by Rudy Krolopp, will become available to consumers next year, priced at $3,995. See 1989, 1998

09-18

09-19

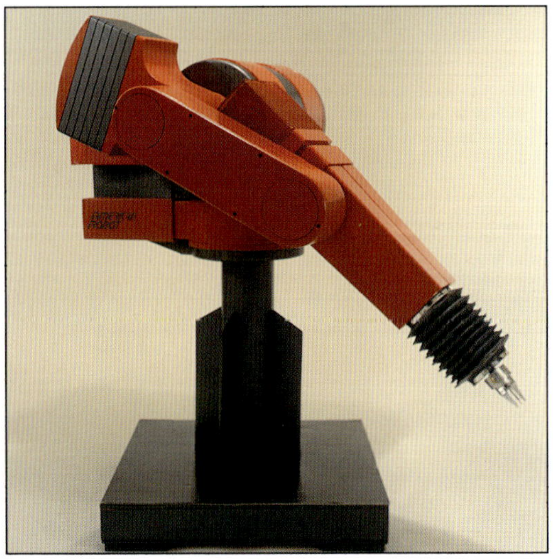

09-20

1984

The CD-ROM, a compact disk (see 1982) that can store very large amounts of digital data and be used with computers, is developed by Philips Electronics, and jointly introduced by Philips and Sony in the US. Sony debuts the first portable CD player, the D-50.

At IDSA's fourth annual IDEA awards program in Seattle, chaired by Read Viemeister, awards include: the Apple IIc *Macintosh®* computer (see illustration); the 1984 Pontiac *Fiero*; the ESPRIT sales booth; the *Merlin* robot (see illustration); the NCR Corporate Strategy; the *Rubber Chair*; and Ciba *Magic* packaging (see illustration).

The Product of Design, an outstanding exhibition that explores the industrial design process, is curated by Joan Wessel and Nada Westerman at the Katonah (NewYork) Gallery includes this quotation by Charles Eames: "The design is an expression of the purpose. It may, if it is good enough, later be judged as art."

Product semantics, the concept of product form as an expression of emotional and functional associations, makes its debut with articles by Klaus Krippendorff, Uri Friedländer, John Rheinfrank, and Reinhart Butter. Co-director of Cranbrook Academy of Art Michael McCoy leads a seminar on the subject entitled, *The New Meaning of Form and Function in Industrial Design.*

The Industrial Society of America (IDSA) and the National Association of Schools of Art and Design (NASAD) announce the commencement of the formal accreditation of industrial design educational programs. An IDSA committee chaired by James Pirkl implements the joint agreement between the two organizations.

A comprehensive *NCR Corporation Industrial Design Strategy* that includes not only a graphic design corporate identity, but also guidelines for consistent product character and relationships around the world, is designed by Richardson Smith, Inc. and the NCR Design Staff under Arnold Wasserman, and is implemented by NCR.

Chrysler introduces the "minivan," advertised as the "magic wagon," whose shape combines the boxy rump of a van with the truncated front end of a car.

American Motors, which in 1969 had acquired the Kaiser Jeep Corporation of Toledo (see 1941,1945), introduces the newest *Jeep*, its new *Cherokee XJ 4-door Sport Utility Vehicle.* This car, designed by Robert Nixon and Vincent Geraci working for Richard Teague, is the first of what will become known as "sport utility vehicles," or SUVs.

09-18. Henry J. Rahn and Weymouth Design complete this *Magic* packaging for Ciba Corning. *Magic* will be named by IDSA in 1990 as a *Design of the Decade.*

09-19. Apple Computer, Inc. introduces its *Macintosh IIc®,* with superior "user-friendly" icon graphics. Design of the *Macintosh* is a collaborative effort between *frogdesign* (see 1983) and Apple staff designers. It includes a much-refined mouse design engineered in 1981 by consultants Hovey-Kelley, founded by Stanford University product design graduates Dean Hovey and Dave Kelley. Their firm later evolves into IDEO (see 1991). The *IIc* is cited as one of best designs of the year by *Time* magazine.

09-20. Alex Bally and Charles Kraeuter of Bally Design, Inc., design this *Merlin Intelligent Robot*, a computerized production assembly tool, for the American Robot Company.

09-21

09-22

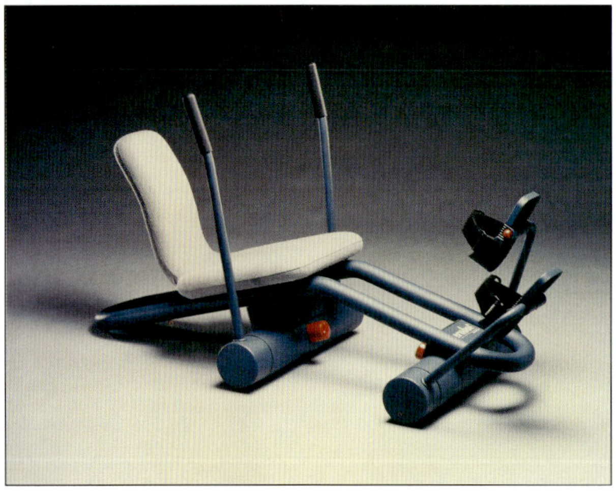

09-23

1985

In their search to produce visible amounts of a theoretical and unstable form of pure carbon with 60 atoms (C60), Robert F. Curl and Richard E. Smalley discover how to produce a stable form, and find that this stability is explained by a molecular structure with the perfect symmetry of a soccer ball. Because the same structural principle underlies the geodesic dome, they name the element *buckminsterfullerene*, or "buckyball" for short, in honor of R. Buckminster Fuller. Their discovery will lead to the discovery of a family of molecules they name *fullerenes*. One of these, to be called *buckytubes*, will be discovered in 1991 by scientist Sumio Iijima.

WORLDESIGN 85, the first Congress of the International Council of Societies of Industrial Design (ICSID—see 1957) to be held in the U.S., is hosted by IDSA in Washington, D.C., and chaired by Deane Richardson. Its *WORLDESIGN* Excellence Center exhibits designs from 93 countries and is attended by 2,300. Registrants vote Charles Eames as the Most Influential Designer of the 20th Century.

At *WORLDESIGN 85*, IDSA presents IDEA awards to the *Metaphor* workstation (see illustration), the *Brigade* firefighter helmet, the *Ethospace* office system, the *Aerobot* exercise machine (see illustration), and the *Facpac* disc storage system.

President Ronald Reagan presents the First Presidential Awards for Design Excellence. A jury selects 13 government projects for the awards, which include the *Unigrid* graphics program, developed by Massimo Vignelli of Vignelli Associates for the National Park Service (NPS); *Unigrid* guided all NPS graphics throughout the 1980s. Vincent Gleason, chief of publications for NPS, leads the excellent program.

High Styles: Twentieth-Century American Design exhibition opens at the Whitney Museum in New York. Assembled by six different curators, it includes 300 pieces of furniture, tableware, and household appliances.

Burroughs Corporation initiates an annual design competition for students, and awards first prize of $10,000 to Larry Vollum, a graduate of California State University, for the design of an interactive electronic music/book stand. Burroughs next year will merge with Sperry Rand, and be renamed the Unisys Corporation; the competition will be continued under that name until the early 1990s.

In Boston, Peter Lawrence establishes the Corporate Design Foundation, which aims to raise the quality of design through business education. He leaves the Design Management Institute (DMI) (see 1974) to the leadership of Earl Powell. See 1994.

09-21. Mike Nuttal, of Matrix Product Design, and Jim Yurchenco, of David Kelly Design, design this *Metaphor Workstation* for Metaphor Computer Systems.

09-22. William T. Dalebout and Sohrab Vossoughi, of Ziba Design, design this *Aerobot* exercise machine for Pro Form.

09-23. The Ford *Taurus* and Mercury *Sable* cars, designed by Jack Telnack, Fritz Mayhew, and Ford staff, are introduced by Ford Chairman Donald E. Peterson, who says he "set the design staff free to create cars that tickled their fancy and in came the *aero look* of the 1980s." Their success will lead U.S. automotive design back to prominence, and the design will be named by IDSA in 1990 as a *Design of the Decade*.

09-24

09-25

09-26

1986

The Machine Age in America 1918-1941, an exhibition curated by Diane Pilgrim for the Brooklyn Museum, opens and tours several U.S. cities. The exhibition catalogue becomes a standard reference work about icons of the early years of industrial design.

Alias, the first practical computer software that allows designers to manipulate form and color in real time, and connect to traditional CAD/CAM systems, is introduced by Alias Research Inc. The computer-aided design program was developed for industrial designers at the request of General Motors.

The death of Raymond Loewy in Monaco is commemorated in a special feature of *Industrial Design* magazine. James Fulton, chairman of the magazine's publisher and a former employee of Loewy, is interviewed on National Public Radio and on several television special features.

At IDSA's annual conference in Chicago, a memorial tribute to George Nelson is presented by Ralph Caplan and Niels Diffrient. IDEA awards are given to the *Vixen Motorhome*, the *Synergy III* sanitation unit, the *Kettle*, and the *Safety-Kleen* container (see illustration).

Color fades from consumer appliances. White is preferred by 68 percent of consumers, according to current surveys, often in "Eurostyle" forms that dominate the market. Second choice for color is Almond at 25 percent, but it is fast declining. The newest "color" in appliances is shiny black.

Alessi introduces the *Bollitore* teakettle with a plastic bird whistle on its spout, designed by architect Michael Graves; it will become one of the most famous kitchenware items of the decade, with one-and-one-half million sold by 2001.

Nike (see 1979) introduces *Air Jordan* athletic shoes designed by Tinker Hatfield and named after basketball star Michael Jordan. Nike now employs a design staff of 160 who use CAD/CAM technology to quickly translate ideas into production.

Winner International markets *The Club*, a bar with a padlock used to lock steering wheels in place and foil car thieves. *The Club* was invented by James Winner, of Pittsburgh, Pennsylvania, and fashioned by Cleveland mechanic Charles Johnson. By 1997, 13 million will be sold at $59.95.

Ford, only half the size of General Motors, tops it in profits for the first time since 1924. The success of the 1986 Ford *Taurus* and Mercury *Sable* design (see 1985) is credited with the turnaround.

Honda Motor Company opens a new design center near Los Angeles, where it had established an office in 1975. Ten other car companies have located their design facilities between Los Angeles and San Diego in the last 13 years. See 1973.

09-24. This Philips Electronic Office System (PEOS), designed by Robert Blaich and Michael McCoy, is awarded a European Ergodesign Prize. It is part of Philips's "Harmonization Program." See 1989.

09-25. Dansk introduces *The Kettle,* an elegant water kettle designed by V. Lorenzo Porcelli. In 1990, Lenox China and Crystal will purchase Dansk.

09-26. This hazardous waste container, designed by George Ira, is introduced by Safety-Kleen. It will be named as a *Design of the Decade* by IDSA in 1990.

09-27

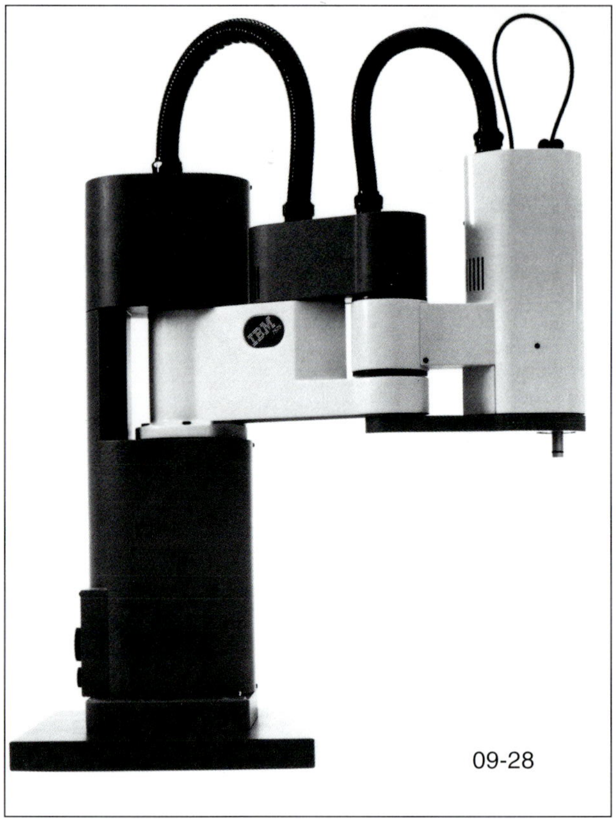

09-28

09-29

09-30

1987

Full-color inkjet printing debuts with the introduction of Hewlett Packard's *PaintJet*. This process will soon replace laser printer technology, which Xerox developed in the early 1980s.

Design in America, an exhibition produced by the USIA, premiers in Belgrade, Yugoslavia, then travels to Zagreb, Ljubljana, and Sarajevo. The display includes more than 250 objects ranging from cars to tableware. The exhibition will also travel to Bulgaria and Hungary in 1988.

Thomson Consumer Electronics, a French government-owned electronics company, acquires the Radio Corporation of America (RCA) from General Electric (GE). Last year, GE acquired RCA, which had been formed and consolidated with GE in 1919 by the government under wartime conditions, but had separated again in 1932. GE was originally formed in 1900 by a consolidation of Edison Electric and the Thomson-Houston Company.

The Dutch Association of Industrial Design (Stechting Industrieel Ontwerpen Nederland—abbreviated as *ioN*) displays 50 products at the 1986 Efficiency Fair that have been awarded the *ioN* designation of "Good Industrial Design." Awards are also given to Philips, Gispen Assenburg, and Eromes, among others.

The National Endowment for the Arts (NEA) (see 1967) premiers *America By Design,* a five-part television series on PBS, which explores the history of American architecture and design.

Krups of Germany presents the first bi-annual Krups Design Awards. First prize of $10,000 is awarded to Marc Harrison, Dov Z. Glucksman, and their team for the design of a personal heater/cooling fan.

At its annual conference, *Monterey 87,* in Monterey, California, IDSA presents ten Industrial Design Excellence Awards (IDEA) to Teledyne speakers, the 1988 *Berretta*, and the IBM industrial robot (see illustration), among others.

Stereolithography is invented and patented by the U.S. firm, 3D Systems. This new method of making prototypes produces plastic replicas of objects, created by computer-aided design (CAD) techniques. A liquid plastic is polymerized by an ultraviolet laser beam that is controlled by a computer, to make the replica that is then lifted out of the liquid as a solid object.

Sunraycer, a solar-powered vehicle designed by General Motors advanced engineering, Hughes Aircraft, and AeroVironment Inc, founded by Paul MacCready, is built at a cost of $8 million. It wins the First International World Solar Challenge 1950 mile race across Australia in November, completing the run in five days at an average speed of 43.5 mph, 600 miles ahead of the nearest of its 25 competitors. See 1977, 1979, 1987.

09-27. The *Hillcrest Motorcaddie*, a remote-controlled, battery-powered golf caddy, designed by Eddie Machen of Machen Montague Inc.; and Mike McCue, president, and Tim Pope, vice-president, of Kangaroo Products Company, debuts.

09-28. Randall W. Martin designs this *IBM 7575* Industrial robot for IBM.

09-29. Marlan Polhemus and Andrew Alger, of Goldsmith Yamasaki Specht Inc., design this *My Way* tape dispenser for Tenex Corporation.

09-30. Kodak's new single-use *Fling* camera repeats history. The entire camera is returned to the maker for processing, as was George Eastman's original 1888 camera. Retail price is $6.95.

09-31

09-32

09-33

09-34

1988

IDSA president Peter H. Wooding and Yuri Soloviev, president of the Society of Soviet Designers (see 1987) sign an "Accord of Mutual Cooperation and Exchange."

Based on supporting evidence submitted by IDSA, the Immigration and Naturalization Service (INS) rules that industrial design is a field of professional endeavor.

Business Week magazine features an illustrated article on industrial design on its cover entitled, "Smart Design: Quality is the New Style." The article will launch a long-term relationship between IDSA's annual IDEA awards and *Business Week*. In the future, the magazine will feature the awards annually, vastly increasing the national visibility of excellent design to the business community.

At *WORLDESIGN 88* in New York in August, IDEA awards presented by IDSA include: the *Newhouse* furniture; the *Voyager* toy (see illustration); the Amoco station; the *Magnum* tractor (see illustration); the Olds *Cutlass*; the *Activity Gym* (see illustration); and the *Spacemaker* (see illustration). Key speakers include Ralph Caplan, former editor of *ID* magazine, and R. David Pittle, Ph. D., technical director of the Consumers Union.

WORLDESIGN Awards are given to three of the world's leading designers; Kenji Ekuan of Japan, Yuri Soloviev of the USSR, and Arthur Pulos of the U.S., in recognition of their lifelong service to the international design community.

The U.S. firm, Scriptel commercializes the first computerized electronic system to enable handwritten graphics to be transferred and stored in computer memory.

Industrial Design magazine changes its name to *International Design* magazine. When a similar change in the profession's name was proposed, the elimination of the word "industrial," only a decade ago, it was unthinkable to the Industrial Designers Society of America (IDSA).

The *NeXT®* computer is dramatically introduced by Steven P. Jobs, who founded *NeXT* in 1985 after he left Apple (see 1977) because of a fierce corporate struggle. The black cube *NeXT* is designed by *frogdesign* (see 1983), and the *NeXT* logo by Paul Rand. The name implies its market objective of being the "next" generation of PCs, with the speed and capacity of a mini-mainframe (A 256-MB optical storage system with 8-MB of RAM) at the low price of $6,500. It will not reach the market until 1991, and will not succeed.

09-31. The *Spacemaker Plus™*, an under kitchen cabinet coffeemaker, can opener, knife sharpener, and task light is designed by Don McCloskey, Sally Hattle and Gary Van Duersen of Black & Decker U.S. Household Group; Group Four Design; and John Howard Industrial Design.

09-32. James Couch, Elizabeth B-N. Sanders, and Keith Kresge, of RichardsonSmith, Inc., design this *Voyager* educational speech recognition toy for Texas Instruments.

09-33. This *4701 Activity Gym,* designed by Kevin Aker, design director for the Little Tikes Company, wins a 1988 IDEA Award. Introduced in 1987, it will be named as a *Design of the Decade* by IDSA in 1990.

09-34. This *Case IH 7100 Magnum Tractor* is designed by Greg Montgomery and staff of Montgomery Design International, and Ralph Lanphere and staff of J. I. Case.

09-35

09-36

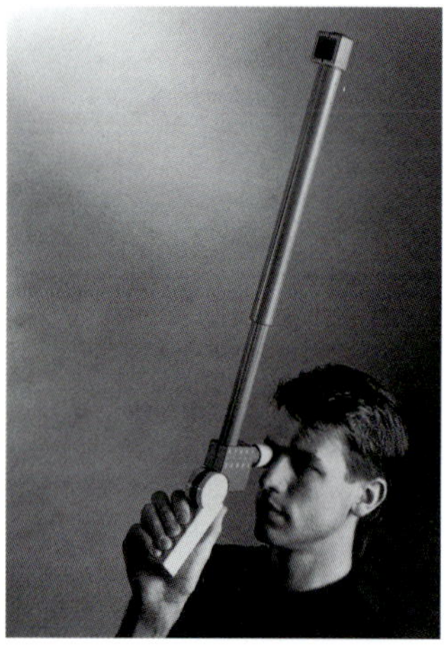

09-37

09-39

1989

Design USA, an exhibition of U.S. industrial design, architecture, and graphic design produced by the USIA, begins a 19-month tour of nine cities in the Soviet Union, beginning in Moscow and ending in Vladivostok. The exhibit will run until February 1991, and will be the last of such government exhibits before the collapse of the Soviet Union in December 1991.

IDEA design awards presented by IDSA at its annual conference *Designing the '90s* in Minneapolis, Minnesota, include *Context* furniture, the *Sportscope* (see illustration), *Cleret* glass wiping tool (see illustration), the *ATFS* footswitch, the Plymouth *Laser* / Mitsubishi *Eclipse*, the *Colorcurve System,* and the TSP *Turret Stockpicker.*

The Design Management Institute (see 1974) partners with the Harvard Business School in initiating a worldwide traveling design management exhibition called TRIAD. See 1985.

Winners of the 1988 Unisys Student Design Competition are announced and include a $10,000 first place award to John Jorgenson of California State University at Long Beach for his *Planchek/Walkabout* construction layout and management tool, and a $6,000 second place award to Terri Capo of the Center for Creative Studies for her *Opti-Scanner* mobile eye screening system.

Adobe introduces its *Photoshop* software for designers that will lead shortly to *Illustrator.* Both computer programs soon become critical tools for graphic designers. Also this year, *T-Flex* and *ProEngineer* 3-D computer-aided design software is introduced, revolutionizing industrial design. These are the first parametric modelers that make the design process intuitive and obvious. Any changes in dimensions or parameters are instantaneously incorporated into the model.

Form Farm, Inc., introduces *Bag Hog,* a bag holder made of a simple plastic sheet and designed by John Lonczak, Tony Baxter, and Simon Yan of Lonczak Design.

Randolph McAusland, president of Design Publications Inc. (see 1978), is appointed as the new director of the Design Arts Program of the National Endowment for the Arts. Within a year, he will be named as Acting Deputy for Programs.

The *Humvee*, built by AM General Corporation to replace the *Jeep* for military services, is put into action during the U.S. invasion of Panama in December. AM General acquired Studebaker's defense business in the early 1950s.The vehicle will later be marketed commercially as the *Hummer*, and become a status symbol, just like its predecessor, the *Jeep* (see 1941). GM will begin marketing the *Hummer* in 2000.

09-38

09-35. *Context* furniture system, designed by the Steelcase Environmental Systems Design Group, is introduced.

09-36. Robert Brunner and Ken Wood, of Lunar Design, design this *Sportscope* Periscope for Carolina Pacific.

09-37. Sohrab Vossoughi, Christopher Alviar, and Paul Furner, of ZIBA Design Inc., design this *Cleret* Glass Cleaner for Hanco. The graceful hand tool is designed to wipe excess water from bathroom shower doors.

09-38. Motorola rolls out this *MicroTAC* cellular phone weighing 10.7 ounces, the world's smallest and lightest, which creates the product category of personal "pocket" telephone. Design is by Rudy Krolopp, Motorola manager of industrial design, along with staff designers Albert Nagele and Leon Soren. By 1996, Motorola's *StarTAC* will weigh a mere 3.1 ounces, and will actually fit in the palm of your hand. See 1983.

09-39. The Mazda *MX-5 Miata*, designed by Mazda staff headed by Tom Matano starting in 1983, is introduced at the Chicago Auto Show. By 1991, Mazda design staff will total 28.

Chapter 10

1990-2000 Integration of Design

At the century's close, design envelops a rich past and a complex present. Industry consolidations compress manufacturing and retailing structures. Giant retailers dictate terms to a decreasing number of manufacturers and rearrange the landscape of how and where products are made and sold. These international economic pressures redefine how designers operate.

U.S. firms outsource manufacturing and, often, design to countries whose lower wages allow more competitive pricing. Many companies establish their own plants in Asian countries. Designers travel around the globe to work with local designers and factories. Design students from China, South Korea, and the Middle East swell U.S. design schools, as those countries respond to the need for design to satisfy international markets. Asian countries establish their own design schools too, often with the assistance of U.S. and European designers.

By the end of the century, design is an integral part of everyday life in a visual culture where even media-savvy children recognize "cool." Design enjoys regular coverage in business magazines, daily newspapers, television features, and a host of industry-supported and professional design awards. Famous architects become even more famous because they design kitchen appliances, and some designers become international celebrities and brands themselves. In a paradoxical consumer culture that spans economic disparities, luxury brands sell for astronomical prices while discount stores set fashion trends in affordable design.

Stylistic influences from all ages and regions blend to suit eclectic tastes. "Retro" styling, borrowing style from the 1930s, 1940s, 1950s, or even the 1970s, becomes popular in automobiles, appliances, and furniture, particularly for young consumers who spot global trends on television and the internet. New synthetic materials bring excitement to the fit, feel, finish, and performance of common products.

"Good" design is now produced by highly creative professionals who are sensitive to user, environmental, multi-generational, and safety issues, as well as economic and performance values. Designers' work methods are light years away from practices of only a decade or two ago. Rapid time-to-market requirements and frequent product line reassessments have become the standards for new product development. Instead of the traditional "artist" tools of sketches, renderings, and polished models, designers now rely on computer visualizations for swift electronic communication with interdisciplinary teams of engineers, toolmakers, materials providers, and product planners around the world.

In addition to designing products, designers provide innovative marketing and strategic planning services for clients. Design firms assist in redefining the consumer experience for products, services, and merchandising environments. Designers have also become anthropologists, as they attempt to discover unmet or unrecognized needs, and their consumer research studies how people truly live with products.

To meet the challenges of the 21st century, designers have expanded beyond their historical roles as they have adapted technology to create products and environments that enhance life for users of all ages and economic levels. They continue to be generalists and specialists simultaneously, and, if recent history is any indication, these talents will continue well into the foreseeable future.

10-01

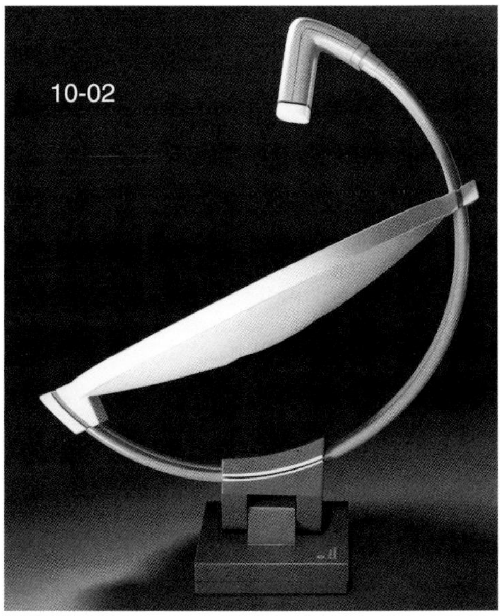
10-02

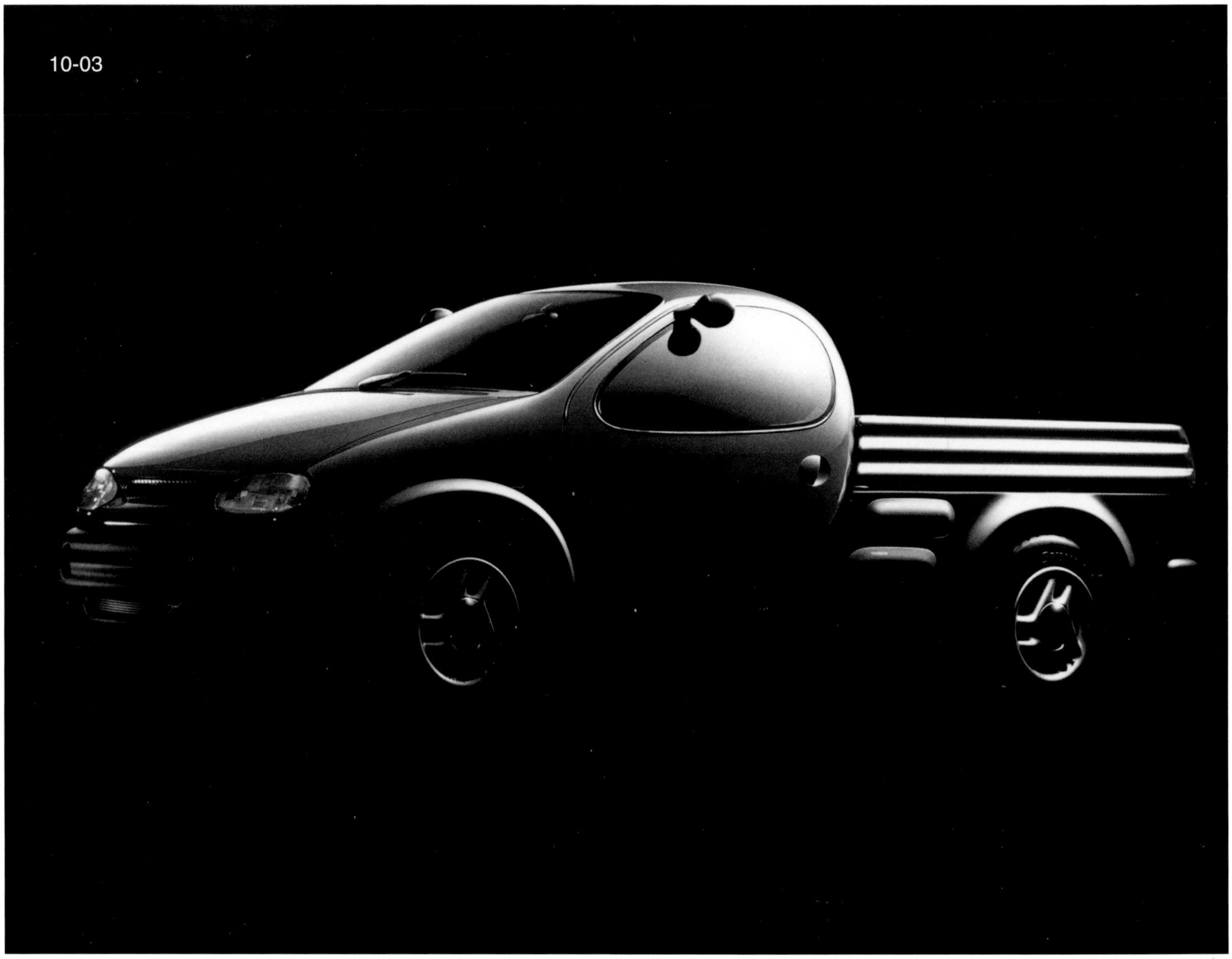
10-03

1990

The appliance market returns to bright colors and accents after years of neutrals. The character of design is also changing from serious and pristine Bauhaus geometry to more whimsical, decorative, and post-modern forms.

At its annual conference in Santa Barbara, California, IDSA presents new levels of Gold, Silver, and Bronze IDEA awards to many, including those to the *Bag Hog* (see 1989), the *AVC* voting machine, a *Driver Assessment System*, the *Gobi* vehicle (see illustration), and a *Satellite Antenna* (see illustration).

Westinghouse announces its purchase of Knoll International and other furniture brands. It forms the Knoll Group, with offices in Pittsburgh and New York. Donald Rorke is appointed executive vice president of design worldwide, to administer the design function for the entire Group.

The American Design Council is formed, comprised of the executive directors of 10 design-related organizations, and will meet semi-annually. Environmentally responsible design was identified as a common priority of all organizations.

Evans & Sutherland announces the availability of its turnkey industrial design software package, CDRS (Conceptual Design and Rendering System). It will radically speed up complex modeling processes, especially in automotive design.

Apple Computer, Inc. appoints Robert Brunner, founder of Lunar Design, as its design director. By 1992 he will build the staff and reduce the hiring of design consultants.

IDSA *Innovation* publishes ten *Designs of the Decade* selected from the previous ten years of IDEA Awards (see 1980) and based on their business success.

Business Week publishes two issues featuring design. Its September special bonus issue, "Innovation-the Global Race," includes an article, "What's Hot in Product Design," and its December article on the IDEA awards, headlined "Rebel with a Cause," features Hartmut Esslinger on its cover. The magazine also announces that beginning next year, it will sponsor the annual IDEA awards program. IDSA will continue to administer the program and select the awards by jury. This will have a dramatic impact on the national visibility of designs and designers.

Windows 3.0, a graphic-based user interface software developed in 1985 by Microsoft for IBM PCs with DOS-based systems following the Macintosh example of mouse and icons, now debuts. *Windows* will soon dominate the home computer market.

The Xerox *Docutech Production Publisher Model 135*, designed by Fitch Richardson/Smith, debuts. It is the first of a digital publishing series for business publishing. In 2000, IDSA will name it a "Design of the Decade."

10-01. *Good Grips* kitchen utensils, with oversized, ergonomic handles, designed by Davin Stowell, Dan Formosa, Tucker Viemeister, Michael Callahan, Steve Russak, and Stephen Allendorf of Smart Design, and Sam Farber (see 1960), Betsey Farber and John Farber of OXO, are introduced by OXO International. In 2000, IDSA will name the line a "Design of the Decade."

10-02. Gregory Fossella Design, a McCalla/Lackey Company; designs this *New Generation Satellite Antenna* for General Electric American Communications.

10-03. The *Gobi* concept vehicle, designed by Gerald Hirshberg, Bruce Campbell and Diane Taraskavage of Nissan Design International, is selected as one of *Time* magazine's annual Best *of Year* designs.

10-04

10-05

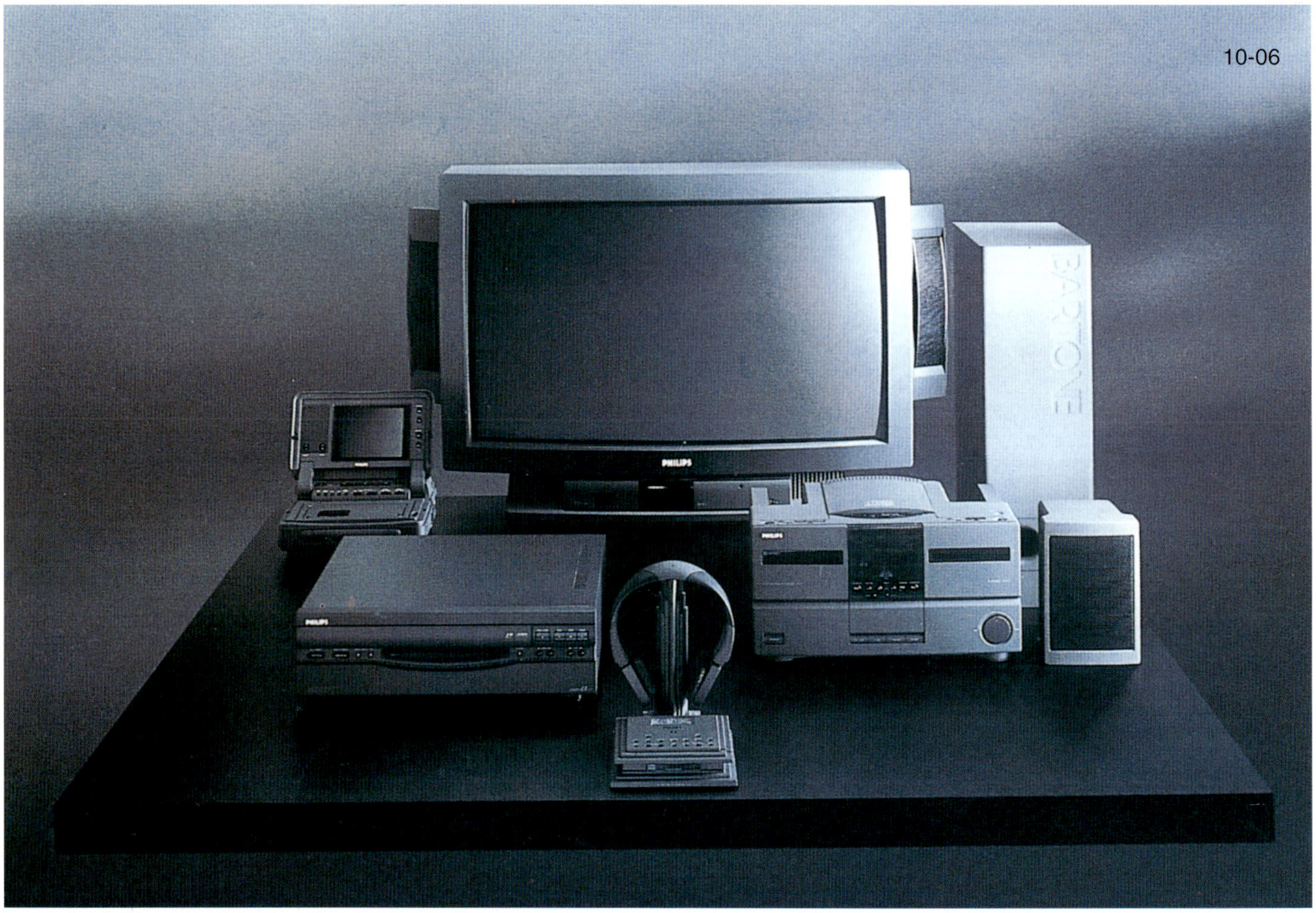

10-06

10-07

1991

For the first time IDEA award winners are featured in a special annual June issue of *Business Week* magazine (see 1990). The awards are presented in August by IDSA at its annual conference, *American Design Revolution/Evolution*, in Boston. Gold award winners include: *ECO 35* camera; *La Croix* packaging (see illustration); Work Group *DEChub* (see illustration); the *Tadpole* (see illustration); and the *FC* truck.

The Japan Industrial Design Promotion Organization (JIDPO) organizes an exhibition, *Good Design (G-mark) Products 90*, in Yokohama City, Japan, that includes participants from winners in the IDEA90 program in the U.S. A highlight is the *Gobi* car/truck provided by Nissan (see illustration 1990).

The Hagley Museum and Library in Wilmington, Delaware hosts a traveling exhibit, *Designing for Product Success*, designed by the Design Management Institute and the Harvard Business School. The exhibit is based on the TRIAD project (see 1989).

Budd Steinhilber and Tom David introduce the concept of "design for disassembly" on a television news broadcast segment on industrial design by Tom Brokaw. A current good example of the concept is the NCR 7731 Personal Image Processor, which is easily dismantled at the end of its life because its parts have been marked for proper sorting and recycling.

The terms "Universal Design" and "Transgenerational Design" are being used to describe concerns for user comfort and safety. The design of products and environments that provide access to the full spectrum of users, regardless of their level of physical ability or disability, is termed "Universal." Products that accommodate young and old, able and less-abled persons are considered "Transgenerational." In the 1970s such concerns were known as "Design for the Handicapped."

Three leading design firms merge to become *IDEO:* David Kelley Design, Matrix Product Design, and ID Two (along with its associate companies in Europe). New principals are David M. Kelley, Bill Moggridge, and Michael Nuttall. By 2004, IDEO will be the largest and most influential product design and development firm in the world, with a staff of 350, an annual revenue of $62 million, and offices in Palo Alto, San Francisco, Chicago, Boston, London, and Munich.

The Apple *Powerbook* ® debuts, designed by Robert Brunner, Gavin Ivester, Susanne Pierce, Jim Halicho, and Eric Takahashi of Apple Computer, Michael Antonczak of Indesign, and Matt Barthelemy of Lunar Design for Apple Computer, Inc. In 2000, IDSA will name it a "Design of the Decade."

10-04. Tony Lane, of The Design Company, designs *La Croix Sparkling Water Packaging* for La Croix Bottled Water Company.

10-05. Digital Equipment Company (DEC) introduces *Work Group DEChub, DECserver 90L & DECbridge 90,* designed by Stuart K. Morgan and Meg Hetfield of DEC.

10-06. Philips Electronics NV introduces the *Philips Collection*, a range of advanced audio and video products that are immediately recognizable as one family because their clean, elegant lines. Design was directed by Robert Blaich and coordinated by A. Labohm.

10-07. This *Tadpole* Infant Positioning System is designed by Michael Gravel of Bissell Healthcare Corporation and Kristine Wohnson of Tumble Forms, Inc.

10-08

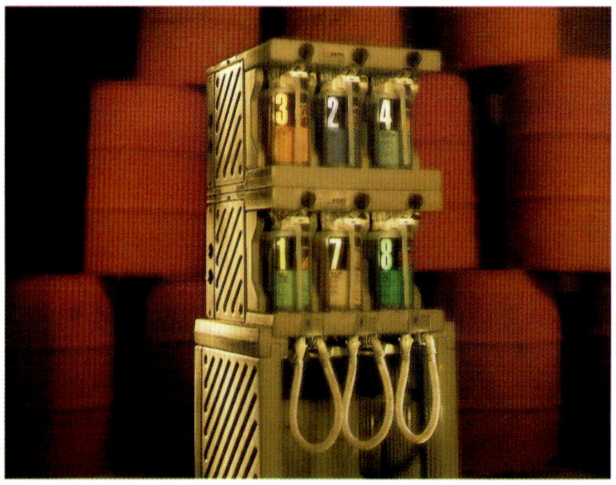

10-09

10-10

10-11

1992

The Americans With Disabilities Act (ADA) is passed by Congress and implemented. It prohibits discrimination on the basis of disability in all places of public accommodation and requires that all new public spaces be designed to be readily accessible to all users.

ICSID holds its biennial congress, *At the Crossroads,* in Ljubljana, Slovenia, Yugoslavia, after an eight-month delay due to political turmoil there. Deane Richardson is elected ICSID's President.

IDEA awards announced in the June *Business Week* magazine and presented by IDSA at its WORLDESIGN 92 annual conference in San Francisco include: the *Aquatred* tire; *Good Grips* (See illustration 1990); the Apple *Powerbook* ®; *CleanWorks* (see illustration); the *AirFLEX* baseball glove; *White to Green* packaging; the GM *Ultralite*; the Cadillac *Seville* (see illustration); and the *Orchestra* lamp. Key speakers at the conference include James Burke, star of the PBS series, *Connections*.

The American Design Council (ADC), comprised of the executive directors of fourteen design-related organizations including IDSA, adopts *The Design Principles of Environmental Stewardship*, prepared by an ADC task force.

Vice President Dan Quayle serves as Honorary Chair of the conference, *Universal Design–Access to Daily Living*, sponsored by Pratt Institute's Center for Advanced Design Research, the Cooper-Hewitt National Design Museum, Smithsonian Institution, and Columbia University's Department of Rehabilitation Medicine.

White to Green Packaging, a simple natural line of corrugated kraft packages are designed by Peter T. Allen, Jean Stevens, Charlie Costantini, Peggy Jensen, Greta Mikkelsen, and Chantale Hansen of Apple Computer, Inc.

Stefano Marzano is named managing director of Corporate Industrial Design (CID) at Philips Electronics NV in Eindhoven, the Netherlands, replacing Robert Blaich, who returns to the U.S. and establishes a design consultancy, Blaich Associates, in Aspen, Colorado.

The Knoll Group debuts the *Orchestra* Desktop Lamp, designed by John Rizzi and Brooks Rorke of RorkeDesign, Ralph Osterhout of Team Machina, and Andrew Cogan and Karl Konrad of Knoll.

Ashlar, Inc. introduces *Vellum 3D,* making on-screen CAD drawing easier.

The *Sensor for Women* razor, with a wide, flat handle and gem-like turquoise concave grip, designed by Jill Shurtleff of Gillette Company design staff, is launched. By 1999 more than 100 million will be sold and, in 2000, IDSA will name it a "Design of the Decade."

10-11a

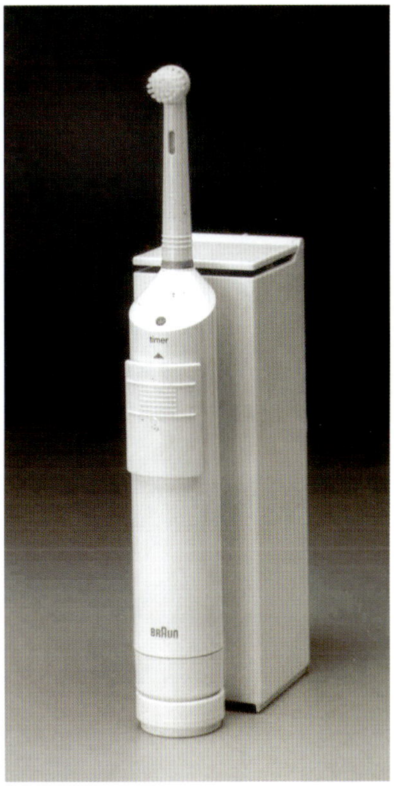

10-08. *CleanWorks,* a dispenser of bulk cleaning chemicals for commercial use, is designed by Alex Bally, Gerald Proctor, Charles Kraeuter, and Doris Wong of Bally Design Inc., and Brandenberg Design and Engineering for Scott/Sani-Fresh Intl. Inc.

10-09. *Personal Harbor Workspace*, designed by Mark Baloga, Paul Siebert and Steve Eriksson of Steelcase Inc. with Greg Draudt and Michael Tingley of Robert Luchetti Associates, debuts. See IDEA Award, 1995.

10-10. Charles M. Jordan, John R. Schinella, Dennis Little and Marv Fisher, of General Motors Design, design this new Cadillac *Seville* STS for Cadillac Motor Division, General Motors Corporation.

10-11. The Goodyear Tire and Rubber Co. introduces this *Aquatred* tire, with an innovative center channel that disperses surface water, designed by Paul B. Maxwell, Sam P. Landers, John Attenello, William Glover, Norm Anderson, Jim Stroble, Don Vera, Joe Hubbell, and William Egan of Goodyear.

10-11a. Braun AG introduces this D 5, its first oscillating toothbrush, designed by Peter Hartwein.

10-13

10-12

10-14

1993

An advisor to president-elect Bill Clinton invites a roundtable of U.S. designers assembled by *International Design* magazine to Little Rock, Arkansas, to "help government restore economic leadership and rebuild our cities." Panel recommendations include the creation of a National Design Council. In 1994, after intensive and exhaustive efforts, and testimony by NEA and IDSA, a bill to establish such a council as an advisory committee to the Department of Commerce will become moot due to an Executive Order by now-President Clinton "restricting the number of future councils and committees."

IDEA awards featured in the June *Business Week* magazine and presented by IDSA at its *Atlanta93* conference, include: the *Color Classic*; a portable cooker; the *3495 Dataserver*; *Softouch* scissors (see illustration); *My First Sony* (see illustration); *Powerbook Duo*; the *Thermal Grill* (see illustration); the *Quest* minivan; The *Trio* vacuum cleaner; and the Ford *Probe.*

The U.S. Information Agency (USIA) presents *Good Design is Good Business,* an exhibition of U.S. products, in Tallinn, Estonia.

The Chicago Athenaeum Museum of Architecture and Design holds its first "Good Design" competition. The Museum of Modern Art presented exhibitions with that same name from 1950 to 1955. This year's theme is *Made in the USA*.

The Organization of Black Designers (OBD) holds its first conference, *Dogon to Digital: Design Force 2000.* The organization was founded by David H. Rice in 1975 and re-organized in 1991 with a membership of 250.

Sears, Roebuck introduces *Robo-Grip* self-adjusting pliers. Invented by William Warheit, a retired engineer, they are improved and brought to production by Dr. Hal Wrigley, a Zelienople, Pennsylvania dentist, who founds Applied Concepts, Inc., to do so. Industrial design is by Bally Design, Inc.

Revere introduces *Excel* cookware, designed by Eric Chan, Jeff Miller, and Eyal Eliav of ECCO Design Inc., to update Revere's classic product line, designed by W. Archibald Weldon and the first cookware to use plastic handles. It was originally introduced in 1939.

Black & Decker and the *PowerShot* Tool Company introduce the *PowerShot* staple gun, designed by IDI/Innovations & Development, Inc. and WorkTools, Inc. In 2000, IDSA will name it a "Design of the Decade."

The Nike *Air Max* running shoe is introduced. It will be selected by *Time* magazine as one of the best designs of 1995.

The AMD-103, designed by Cesar Vergara and the Amtrak staff, is the first U.S. passenger locomotive in 30 years to involve an industrial designer (see BART, 1972). It is produced by General Electric. See 2000.

10-12. Doug Birkholz, Craig Melter, Steve Ruelle, and Paul Hendon of Fiskars, Inc., design these *Softouch* Scissors.

10-13. *My First Sony*-Electronic Sketch Pad and Animation Computer, designed by Sony Design Center for Sony Corporation as extensions of the *My First Sony* line initiated in 1987.

10-14. The *Thermal Electric Grill,* designed by Dale Benedict, Susan Haller, Greg Breiding, Mark Ciesko, Andreas Roessner, and Inars Jurjans of Fitch, Inc., and Monte Peterson, Rich Bertelli, Jim Fitzgerald, David Rickords, Fred Mather and Pat Francomb of the Thermos Company, for Thermos.

10-15

10-16

10-17

1994

President Bill Clinton presents eight Presidential Awards for Design Excellence. Among the awards for architecture and graphic design is one award for an industrial design, the hand-held EGIS Explosive Detector designed in part by Design Continuum.

The British Design Council (see 1944) publishes its last issue of *Design* magazine, which began publication in 1949. See 1995.

Canada's Prime Minister opens the Design Exchange (DX), a private institution housed in the former Toronto Stock Exchange and chaired by Donald Macdonald. Representing architecture, industrial design, graphic design, fashion design, and landscape architecture, DX aims to increase design awareness and innovation as tools for economic growth. See Design Canada, 1961, 1976.

IDEA awards, published by *Business Week* magazine in June and presented in Detroit by IDSA in August, include: *Kin-der-link* furniture (see illustration); the *Zephyr* hammock; the *Newton Message Pad*; *Vision Date +* camera; *PowerShot* (see 1993); and *The Measure of Man and Woman*.

The National Housewares Manufacturers Association (NHMA) initiates an annual *Design Oasis* exhibition at its International Housewares Show, as well as an annual national student design competition. Victoria Matranga administers both programs for NHMA. After 2003, the organization will be known as the International Housewares Association (IHA).

Among the winners of the second annual Chrysler Award for Innovation in Design are Katherine and Michael McCoy, for their influential work on the symbolic and semantic meaning of design.

In San Francisco, Kelly Hand, Tom Jacobson, and Wendy Sommers found the Association of Professional Model Makers (APMM).

The Council of Federal Interior Designers, Institute of Business Designers, and International Society of Interior Designers merge to form the International Interior Design Association (IIDA), representing 8000 members.

The Measure of Man and Woman, a human factors design reference book compiled by Alvin Tilley, Rebecca G. Welles, James Ryan, and John McFarlane of Henry Dreyfuss Associates, is published.

Van Nostrand Reinhold publishes *Transgenerational Design: Products for an Aging Population*, by James J. Pirkl, former chair of the department of design at Syracuse University. (see 1991).

The Volkswagen *Concept One* "retro-Beetle," designed by J Mays and Freeman Thomas under Charles Ellwood, of the VW /Audi studio in Simi Valley, California, debuts at the NY Auto Show. It will come to market in 1998.

Chrysler's *Neon,* designed by Thomas C. Gale and Chrysler's design team, enters the market. Chrysler's purple Plymouth *Prowler* "drag-racer"-style concept car is first shown and will also be produced (see 1997).

Detroit rebounds to recapture 75 percent of U.S. automobile market share, and, for the first time in 14 years, leads the world in production. The Big Three automakers reach their all-time record earning levels.

10-15. The *Snakelight* ™, designed by Richard Carbone and Brian deBlois of staff design, is introduced by Black & Decker, U.S., Inc.

10-16. The *Double Colander,* designed by Morison S. Cousins, Tupperware's vice president of design, is released by Tupperware Worldwide.

10-17. Bashir Zivari of Skools, Scott Ross of Autumn House, and Jason Voyce of Marshall Group design this *Kin-der-link Furniture for Kids* for Skools, Inc.

10-18

10-19

10-20

1995

Toy Story, a full-length Disney computer-animated film by John Lasseter and the first created wholly by computer, is produced by Pixar Animation Studios and becomes an instant hit. Steven Jobs (see 1977) bought Pixar in 1986 for $100,000. Several industrial designers participate in the film's production to insure realism in the appearance of manufacturing details in the molded plastic toy characters.

An exhibition, *Designing Modernity–The Arts of Reform and Persuasion, 1885-1945,* opens at the Wolfsonian Foundation in Miami Beach. Curated by Wendy Kaplan, the exhibition presents items from the collection of Michael Wolfson, Jr., of Miami Beach, Florida, founder of the Wolfsonian Design Museum. The exhibition will travel to Los Angeles, Seattle, Pittsburgh, and Indianapolis, and the catalogue will become a valuable reference work.

The British Design Council (see 1944) is reinvented to "play a major role in the renewal and regeneration of UK industry," according to Council chair John Sorrell. Its magazine *Design*, recently discontinued (see 1994), is now reintroduced.

NEA chair Jane Alexander presents 77 Federal Design Achievement Awards, which include the *60K Loader Cab* interior by Fitch, Inc., and the *Backpack Personal Cooling System* designed by Carlson Technology Inc. In November, NEA's budget is cut 40 percent to $99 million.

IDEA awards are featured in the June *Business Week* magazine and presented by IDSA at its September conference, *Natural Resources*, chaired by Tucker Viemeister in Santa Fe, New Mexico. Winners include: the *Snakelight* (See illustration 1994); the *Jeep* Stereo System (see illustration); Steelcase's *Personal Harbor Workspace* (see illustration 1992); the *Speedray* kickboard; DeWalt's drill /screwdriver (see illustration); and Chrysler's minivan.

The Chicago design firm of Goldsmith Yamasaki Specht (GYS) is sold to First Company, a design office in Kansas City, Kansas. Paul Specht was the last remaining original partner in the firm. David McCormick, partner of First Company, becomes president of GYS and will retain the Chicago office until 1997, when it will be sold to Sundberg Ferar. GYS will close its Kansas City office in 2001; the Chicago office will close in 2002.

Garbino (or Garbo), an inexpensive curving translucent plastic wastebasket designed by Karim Rashid for Umbra, debuts. It will become a signature product for Umbra, and Rashid will be named Young Designer of the Year in 1998 by the Brooklyn Museum.

Chris Bangle, previously with Opel and Fiat, is named as director of design for BMW, the first American to hold this position. BMW purchases a controlling interest in *Designworks* of Newbury Park, California, for design support.

10-18. Skip Kirk, Paul Lacotta and Walter Stoeckmann, of Human Factors Industrial Design, design this *Jeep* portable Stereo System for Kash 'N Gold.

10-19. Robert Somers and Donald Zurwelle, of Black and Decker's design staff, design this DeWalt *Pistol Grip Drill & Screwdriver.*

10-20. This ergonomic *Aeron* chair, designed by Bill Stumpf of Stumpf, Weber + Associates and Don Chadwick of Chadwick & Associates, is launched by Herman Miller, Inc. In 2000, IDSA will name it a "Design of the Decade."

10-21

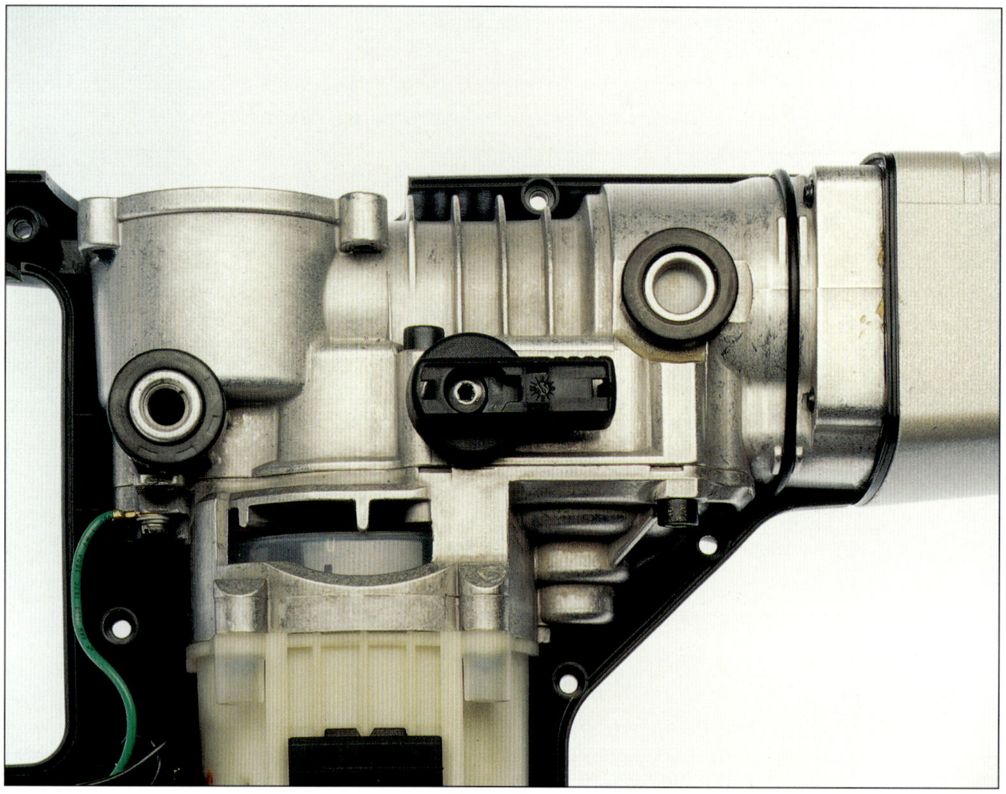

10-22

10-23

1996

The U.S. Bureau of the Census establishes Standard Industrial Classification Codes for design, spurring the inclusion of industrial design, graphic design, and interior design in the 1997 Economic Census. The applicable code is 54142, "Industrial Design Services," and applies in the U.S., Canada, and Mexico.

IDEA awards announced in the June *Business Week* magazine and presented in September at the IDSA national conference, *Worldesign '96–Alternative Realities,* chaired by Charles Allen in Lake Buena Vista, Florida. They include the Microsoft *EasyBall* (see illustration); The Black & Decker *Macho* Rotary Hammer (see illustration); the *Solstice* umbrellas (see illustration); Rock and Roll interactive multimedia; and the *Sea-Doo* jet boat.

The IDSA conference keynote speaker is Michael Graves, architect of the whimsical Walt Disney World Dolphin and Swan resort hotels. Graves begins his lecture with "Mies van der Rohe said that 'less is more.' At Disney, 'more is not enough!'"

The Electronics Industries Association (EIA) initiates a design awards program, *Innovations '96*, offered in conjunction with its Consumer Electronics Show (CES) beginning with the January 1996 show. Entries from 323 winners are displayed and viewed by 30,000 attendees.

The National Housewares Manufacturers Association (NHMA) conducts its annual Student Design Competition (see 1994), awarding $11,000 in cash prizes. Tied first-place winners are Jennifer Nemec (University of Illinois at Chicago) and Charles Loeb-Munson (Cleveland Institute of Art), with product concepts that reflect the healthy, cook-at-home needs of the American family. Nemec designed the "Noodlini" children's pasta maker and Loeb-Munson designed the "Pizza-Press" pizza maker.

Sony introduces its *Playstation*, a product it describes as the most important product since the *Walkman*, to compete with *Nintendo* and *Sega*. Designed by Ken Kutaragi, the product uses 3D graphics, "real time" on-screen action, and CD-quality sound. It is powered by five processors computing half a billion instructions per second.

The *PalmPilot 1000*, a shirt-pocket-sized computer used for tracking appointments, addresses, and phone numbers, debuts. Designed by Palm Computing's chief technologist Jeff Hawkins, its sales will quickly soar to one million units. In 2000, IDSA will name it a "Design of the Decade."

Canon introduces the *Ixus*, a camera the size of a pack of playing cards, which is one of the smallest cameras in the world. Designed by Canon staff, it is hailed as a design classic. The camera uses a new type of 24-mm wide film in an easy-load cartridge, the Advanced Photo System (APS).

10-21. Black & Decker (B & D) debuts a new global range of rotary hammers, designed by Donald W. Zurwelle, manager for professional power tools of B & D, U.S., and Michael Stirm of B&D, Germany.

10-22. Microsoft introduces this *EasyBall* computer input device, a bright, oversized trackball that is easier for children to use than the traditional mouse. It is designed by Fernd van Engelen and James McCay of Walter Dorwin Teague Associates, and Edie Adams, Christopher Alviar and Steven Kaneko of Microsoft Corporation.

10-23. This *Solstice Collection* of patio umbrellas, designed by architect Robert Chipman of Austin, Texas, to withstand 85 m.p.h. winds, is introduced by Landscape Forms, Inc., of Kalamazoo, Michigan.

10-24

10-25

10-26

1997

A *Time* magazine article, "Where the Jobs Are," predicts "76,000 new product design jobs by 2005," and anoints design as "one of the hottest professions in terms of job growth, and compensation." The forecast is based on figures from the Bureau of Labor Statistics, which states that industrial designers (current total—9,000) are among these design-related professions and are projected to reach 21,000 by 2005.

IDEA awards announced in a June *Business Week* magazine and presented in August at the IDSA national conference, *Blurring Boundaries,* chaired by Bob Brunner in Washington, D.C., include the Boston stapler (see illustration), Canon's *ELPH* (see illustration), Perfect Fit Skates, and Ford's *Ka* (see illustration).

The ICSID '97 Congress is held in Toronto, Canada, with the theme. "The Humane Village," Augusto Morello of Italy is elected ICSID president.

The exhibition, *Henry Dreyfuss Directing Design: The Industrial Designer and his Work*, opens at the Cooper-Hewitt National Design Museum, Smithsonian Institution, New York. Concurrently, Rizzoli International publishes, *Henry Dreyfuss, Industrial Designer: The Man in the Brown Suit*, by Russell Flinchum.

Designer, educator, and entrepreneur George M. Beylerian establishes *Material ConneXion*, a designers' resource for innovative materials, in New York, with support from Steelcase, Inc.

Perfect-Fit inline skates, with precise sizing and *QuickClick* detachable frames, are introduced by Launch Sports, Chanhassen, Minnesota, and designed by Rick Polk, Adam Luedke, Jim Easley, Ken Schomberg, and Scott Olson, the founder of Rollerblade, all of Leisure Inc. of Chanhassen.

Indiana real estate developer Chris Baker attempts to revive production of the *Avanti* (see 1962), to be named *Avanti III.* He engages Tom Kellogg as chief designer for the intended 1998 models.

A purple Plymouth *Prowler* "drag-racer" style car (see 1994) debuts at a price of $35,000, but, by 1998, limited production and high demand will escalate bids into the $100,000 range. In 1999, Chrysler Corporation will retire the Plymouth name.

Mercedes Benz debuts a concept car named the *Maybach* (after Wilhelm Maybach, who in 1896 built cars with Gottlieb Daimler), a 12-cylinder super luxury car to compete with Rolls Royce and Bentley. The production versions (Models 57 and 62) will be introduced in 2004, with design credits to Professor Gaus and Stephen Mattin. Price tag: $300,000 to $360,000.

10-24. This Boston desk-top stapler, the first vertical-standing configuration, is introduced by the Hunt Manufacturing Company in Philadelphia, and designed by Bob Brunsdon, senior designer, and George F.W. Boesel, president and owner, both of Spectrum Boston Consulting, Boston, Massachusetts.

10-25. Canon Inc. targets young independent women with the introduction of its *ELPH*, a compact camera using its new 24-mm Advanced Photo System (APS). It is designed by Yasushi Shiotani of Canon, Inc., Tokyo.

10-26. Ford's *Ka,* based on a concept car revealed at the 1994 Geneva, Switzerland Motor Show, debuts. Ford's first entry into the Sub-B small car class, it was designed by Jack Telnack, Claude Lobo, Fritz Mayhew, Robert Giachin, Chris Svenssen, Pierre Webster, Simon Bury, Shareen Rapaport, and Barbara Scheffler of Ford Corporate Design. J Mays (see 1994) is named Corporate vice-president of design, replacing Jack Telnack.

10-27

10-28

10-29

1998

The Art of the Motorcycle exhibit at the Guggenheim Museum in New York resembles a parking-garage ramp and is the novelty blockbuster of the year.

Sony, RCA, Panasonic, and other makers bring high-definition TV sets to the market. With up to 1,080 lines of resolution, (compared to 330 or so on conventional sets), the televisions offer spectacular images, but retail at $7,000 and up.

The first portable MP3 players are introduced by Rio, Diamond and Multimedia.

The first-ever Medical Design Excellence Awards (MDEA) debut, sponsored by Canon Communications, and administered and endorsed by IDSA. Jury chair Kent Ritzel announces the awards, which include: *The Wand*, a dental anesthesia delivery system designed by Herbst LaZar Bell for Milestone Scientific; the *CombiCarrier*, a stabilizing device for transporting patients to the hospital, designed by Morphix Design for Hartwell Medical Corporation; and *Temporary Pacemakers*, handheld versions designed by IDEO Product Development for Medtronics, Inc.

Announced in June *Business Week* magazine and presented at its annual conference in San Diego, with the theme "Why Design? Why not?" chaired by Ken Schory, IDSA's 1998 Industrial Design Excellence Awards include Motorola's *Talkabout* (see Illustration), Sun Microsystem's *JavaStation* (see illustration), Black & Decker's *Kitchentools*™, the *Music Gismo* (see illustration), and the Volkswagen *Beetle*.

Black & Decker (B & D) introduces its *Kitchentools*™ line of appliances that include a can opener, portable and stand mixers, a blender, a coffeemaker, and a food processor. Design is by Mike Laude and Stuart Naft of B & D, and Jan Hippen, Iulius Lucaci, Angelika Schubert-Belle, and Bob Sweet of ZIBA Design.

The long-awaited new *Beetle* (see 1994) arrives and is priced at $18,000, but black market boosts some sales to over $25,000. The new *Beetle* has AC and front wheel drive. Design is by Design Center California of Volkswagen of America, Simi Valley, California, and Volkswagen Design Center of Volkswagen, Wolfsburg, Germany. In 2000, IDSA will name it a "Design of the Decade."

Daimler-Benz of Germany, parent company of Mercedes, acquires Chrysler for $39 billion, the largest takeover of an American company by a non-U.S. corporation in history. The new company is named *DaimlerChrysler*, now the fifth-largest automotive maker in the world. Tom Gale is design chief.

The *Smart* (for Swatch-Mercedes ART) vehicle, featuring changeable plastic body panels to customize color schemes, is introduced by Mercedes, based on its *Swatchmobile* concept car of 1994. The car is styled by SMH Automotive of Switzerland. Mercedes also launches its new ML320, the first luxury class SUV, powered by a V6 engine and assembled in Alabama.

10-27. Motorola introduces this *Talkabout* two-way radio, designed by Glen Oross, Frank M. Tyneski, Michael Page, and Bruce Claxton, of Motorola staff. In 2000, IDSA will name it a "Design of the Decade." See 1983, 1989.

10-28. Sun Microsystems releases its *JavaStation*, hardware capitalizing on its web-friendly programming language, *Java,* introduced in 1996. Designers include Philip Yurkonis, Craig Leverault, Rob Lajara, Kuni Masuda, Cliff Willis, Alison Armstrong, Craig Hartley, and Jean Lynch of Sun Microsystems; and Paul Montgomery, Herbie Pfeifer, Ted Bryant, and Alex Fekete of Montgomery/Pfeifer, San Francisco.

10-29. The *Music Gizmo*, a toy designed by Klitsner Industrial Design, debuts.

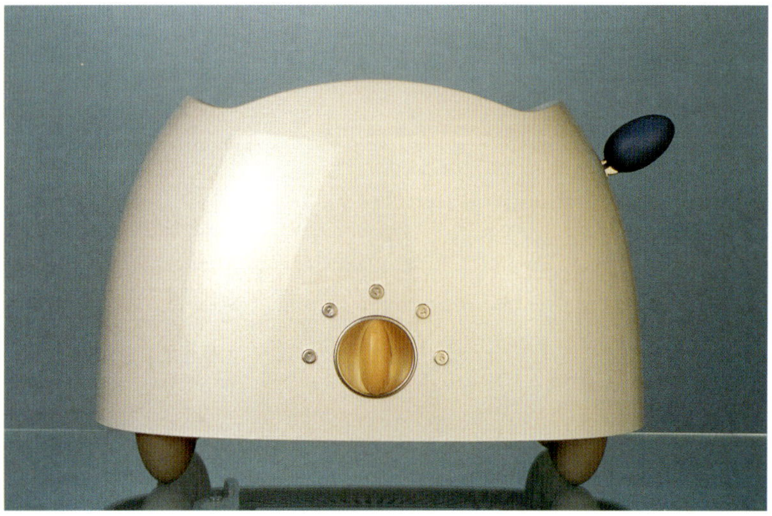

10-30

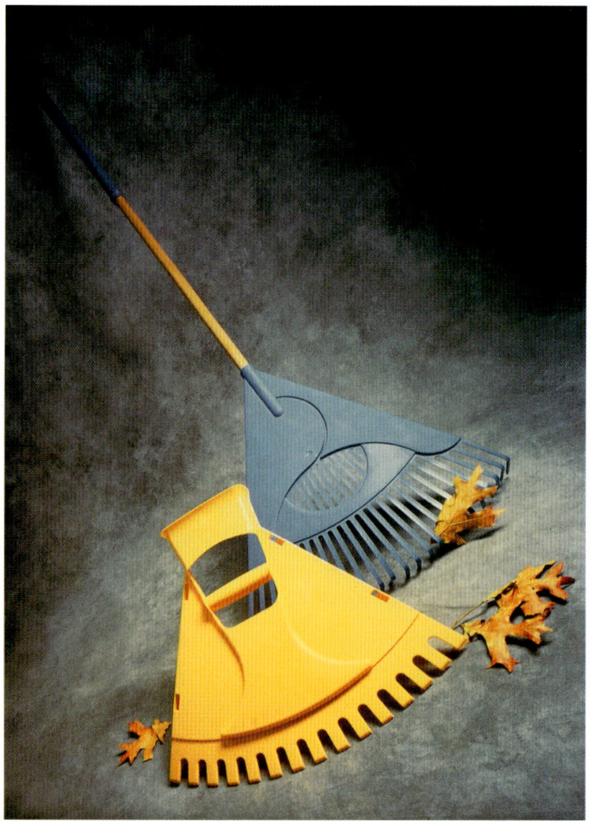

10-31

10-32

10-32a

1999

At its annual conference in Chicago, IDSA confers IDEA awards, which include the *iMac* ® (see illustration), The Apple *Power Mac* ® *G3*, Apple's *Powerbook* ® *G3*, the OXO *Good Grips* Salad Spinner, *Gather & Go Rake* (see illustration), and the Target Toaster (see illustration). Awards are published in *Business Week* magazine in June.

In December, *Business Week* magazine sponsors and publishes winners of *The Designs of the Decade: Best in Business 1990-1999 Awards* competition, coordinated by IDSA. Winners include the *Aeron Chair* (see illustration 1995), OXO *Good Grips* Kitchen Tools (see illustration 1990), *PowerShot* staple gun (see 1993), the new VW *Beetle* (see1998), Apple's *iMac* ® (see illustration 1998), Motorola's *TalkAbout* radio (see illustration 1998), IBM's *ThinkPad*, Apple *PowerBook* ® (see 1992), Xerox's *DocuTech* (see 1990), AVC voting machine (see 1990), Gillette's *Sensor for Women* (see 1992), and *Sea-Doo* watercraft (see 1996).

In Philadelphia, the Philadelphia Chapter of the American Institute of Architects founds the Charter High School for Architecture + Design (CHAD). It is a public, tuition-free school that employs design and architecture to engage and empower students.

Target stores launch a new 150-item line designed by architect Michael Graves, which includes a new teakettle, toaster, peeler, garlic press, can opener, and spatula. The products are produced by Black & Decker and Mirro, among others. The toaster (see illustration) premiers the line.

Nokia's 8860 cell phone, designed by Nokia's chief designer, Frank Nuovo, debuts. Nokia surpassed Motorola in sales last year, due to its 1994 design innovations that transformed the cell phone from a communications tool to a fashion statement.

The North American International Auto Show bristles with "retro" designs. *DaimlerChrysler* offers its Chrysler *PT* (for Personal Transportation) *Cruiser* concept/production car, with full fenders and running board, designed by Bryan Nesbitt, Jeffrey Godshall, Steve Ferrerio, Mike Nicholas, John Ozark, and Eli Nichols of *DaimlerChrysler* Corporation. Ford rolls out a *Thunderbird* designed by J Mays, due in June, 2002.

Sport Utility Vehicles, popularly known as SUV's, dominate automotive sales. Sales have increased from 50,000 to nearly 160,000 annually in the last seven years. Designed on a heavy truck platform with generous cargo space, SUVs appeal to consumers for crash survivability, road visibility, and utility.

10-30. This toaster is first in the line of Target products. It is designed by Michael Graves, Donald Strum, and Tony Boczkowski of Michael Graves & Associates.

10-31. True Temper Hardware launches this *Gather & Go Rake*, designed by John Keily, Keith Kristiansen and Gary Grossman of Innovations and Development, Inc., and Henry Daniels of True Temper Hardware. A second, detachable head forms a clamshell to pick up leaves.

10-32. This Apple *iMac* ® computer in translucent colors, is designed by Bart Andre, Danny Coster, Danielle De Iuliis, Richard Howarth, Doug Satzger, Cal Seid, and Christopher Stringer of Apple staff under British-born Apple industrial design head Jonathan Ive (b. 1967), who joined Apple in 1992, and Steve Jobs, who rejoined Apple in 1996 as CEO. The *iMac* will become the fastest-selling computer in Apple's history and in 2000, IDSA will name it a "Design of the Decade."

10-32a. *The Freedom Chair*, designed by Niels Diffrient and made by Humanscale, Inc., is introduced.

10-34

10-33

10-35

10-36

2000

The Cooper-Hewitt, National Design Museum, Smithsonian Institution, holds *Design Culture Now,* an interdisciplinary conference in conjunction with the Industrial Designers Society of America and the American Center for Design. In November, the Cooper-Hewitt announces winners in its inaugural National Design Awards. Lifetime achievement awards go to architect Frank Gehry for his Guggenheim Museum in Bilbao, Spain, to Apple Computer, Inc. for corporate achievement, and to inventor Paul MacCready for exemplary work in the design of technology.

Women Designers in the USA, 1900-2000: Diversity and Difference, an exhibit at the Bard Graduate Center for Studies in the Decorative Arts, presents the work of many unrecognized women, as well as work by Florence Knoll, Katherine McCoy, Nancy Perkins, Jill Shurtleff, Lisa Krohn, and Lella Vignelli.

The Metropolitan Museum of Art opens an exhibition, *American Modern 1925-1940 Design for a New Age*. It will travel the country to five other museums through 2002.

At the International Housewares Show, the seventh National Student Design Competition (see 1994) includes 212 entries from 27 U.S. design schools. First place, and $3,000, goes to Manuel O. Saez, a senior at the University of Bridgeport, for his Daisy Air Purifier.

The March 20 issue of *Time* magazine includes a feature article illustrated on the cover, *The Rebirth of Design*, sub-titled, *The Redesigning of America*. The article declares "function is out, form is in" and focuses on the "design economy," industrial designers, and many current examples of product design.

At the IDSA annual conference in New Orleans, Louisiana, annual IDEA awards are presented, including Apple's *iSub* ® (see illustration) and *Cinema Display*, *Ojex* Juicer (see illustration), Tupperware cereal server, *Quick'n Easy* food processor, Delta faucet, and Tenex *X-Files* (see illustration).

The first mass-produced gasoline-electric hybrid vehicles, the Toyota *Prius* and the Honda *Insight*, enter the market. GM unveils its *Precept*, a test car that can travel 80 miles per gallon.

A wave of new, upscale scooters based on skateboard technology, which began in Japan last year, reaches the U.S. and is led by the best-selling two-wheeled *Razor* brand.

Amtrak unveils its new high-speed (150-mph) train, *Acela*, with service to begin next year. The train will reduce travel time between Boston and New York to three hours, and between New York and Washington to two-and-one-half hours. Design is by Aura Oslapas, Nick Oakley, Ilya Prokopoff, Johnson Chow, Alycia Freeman, and Richard Eisermann of IDEO; Nikolas Frank, Adrian Corry, Brent Oppenheimer, Robert Haueter, Mary Ellen Buttner, Robert Homack, David Shields, Jonathan Soard, Ginger Strand, Nancy Wolff, and Stuart Henley of OH&CO; and Donald Khapik and Richard Sarles of Amtrak/National Railroad Passenger Corporation. See 1993.

10-33. Tenex *X-Files*, designed by Mark D. Dziersk, Jose M. Perez, and Ken Bargo of Herbst LaZar Bell, and Albert Cheris of Tenex Corporation, for Tenex, debut.

10-34. Anthony Di Bitonto, Marco Perry, Tim Kennedy, and Alex Hochstrasser, of Smart Design, design this *Ojex* Manual Citrus Juicer for OrangeX.

10-35. Harmon/Kardon *iSub* ® sub-woofer for *iMac* ® and the 22-inch Apple Cinema Display ® designed by Apple Computer, Inc., debut.

10-36. BraunAG introduces this *Syncro* shaver, designed by Roland Ullmann.

Chapter 11

Biographies

Alvar Aalto (1898-1976). Finnish architect and furniture designer/manufacturer who studied architecture at Helsinki Polytechnic, 1916-1921, and opened an office in 1923. After 1929, he associated with László Moholy-Nagy, Georges Braque, and Fernand Léger. Aalto developed a new process for laminating plywood that was shown at an exhibition in Helsinki and led to the establishment of his export company, Artek. His inexpensive chairs humanized International Style by emphasizing organic form. He was featured in a one-man show at the Museum of Modern Art in 1938 that had a major impact on American designers. He was a visiting professor at MIT from 1946 to 1947.

Josef Albers (1888-1976) Born in Westphalia, Germany, he studied art in Berlin, Essen and Munich from 1913 to 1920. He then studied at the Bauhaus, became a teacher there in 1923, and was the first student to become a Bauhaus 'master' in 1925. In 1928, he succeeded Marcel Breuer as head of the Bauhaus furniture workshop until 1933, when he came to the U.S. He taught at Black Mountain College until 1949, and lectured at Harvard and Yale. In 1950 he became professor of color theory and chair of the design department at Yale. In 1963, he published *Interaction of Color.*

Robert F. Andrews (1923-1991) U.S. self-taught automotive designer who grew up in Monroe, Michigan and began his career at the Monroe Auto Equipment Company. In 1945 he joined Hudson Motor Car Company working for Frank Spring and developed Hudson's unique 1948 "step-down" chassis. In 1946 he joined Willys-Overland and worked on the *Jeepster* and *Jeep* station wagon. He joined Raymond Loewy in 1949 and in 1960 became a member of the Loewy team that designed the Studebaker *Avanti.* Andrews later became a contributing editor to *Automobile Quarterly* and *Classic Car* magazines. He retired to Monroe in 1985 and devoted himself to sculpture and painting.

Wallace Henry Appel (b.1925) U.S. industrial designer who was born in Boston, attended Penn State University, and graduated from the Rhode Island School of Design. He joined Westinghouse in 1950, was appointed manager of Consumer Products Design in Columbus, Ohio in 1955, and design manager for Mansfield Center in 1969. After Westinghouse was acquired by White Consolidated Industries (WCI) he was appointed manager of industrial design for Mansfield Products Company of White Westinghouse from 1975 until 1980, then became vice president of design. In 1982 he was named vice president of design for WCI International, a position held until his retirement to Venice, Florida in 1991.

Egmont Arens (1888-1966). A U.S. package and industrial designer, he was sent by his family to New Mexico to recover from tuberculosis and began a career as sports editor for the Albuquerque Tribune-Citizen in 1916. He moved to New York, operated the *Washington Square* bookstore, and established the *Flying Stag Press* to publish magazines. His interest in art led him to editorship of *Creative Arts* magazine, *Playboy* (the first magazine devoted to modern art), and *Vanity Fair,* a leading style and fashion magazine. By 1929 he was advertising director for Calkins & Holden, where he headed an industrial styling department until it was discontinued in 1936. He wrote *Consumer Engineering* with Roy Shelton in 1932, and designed the Higgins inkbottle, A&P packages including 8 O'clock Coffee, the classic *KitchenAid* stand mixer, and the Phillip Morris trademark.

Alfons Bach (1904-1999) U.S. industrial designer, born in Magdeburg, Germany who worked in film set design and studied architecture at the Reimann School and the University of Berlin before immigrating to the U.S. in 1926. In 1932 he established his own firm, Alfons Bach Associates, in New York. Clients included Heywood-Wakefield Furniture Company, Lloyd Manufacturing Company, International Silver, Philco and TWA. He settled in Stamford, Connecticut in 1937, where he designed and developed Ridgeway Shopping Center. He became president of the American Designers Institute (ADI) in 1943, moved to Palm Beach, Florida in 1959, where he continued his practice, and retired to Pensacola, Florida in 1992.

Walter P. Baermann (1903-1972) Design educator who was born in Germany and studied architecture and engineering at the University of Munich. After coming to the U.S., he worked as a designer in New York (1932-1937) and served as educational director of the Springfield, Massachusetts Art Museum. He started a graduate design program at Caltech in 1937, where he stayed until 1940. In 1942 he went to Cranbrook Academy to head the design program. In 1951 he created his own office which moved to Raleigh North Carolina in 1964, where he resumed teaching industrial design at North Carolina State.

James Gilmore Balmer, Jr. (b. 1922) U.S. industrial designer who was born in Pennsylvania and graduated from Carnegie Institute of Technology in 1943. He worked from 1946 to 1959 for Harley Earl Associates, Inc., served in a variety of positions, and ended as executive vice-president. In 1959 he was the co-founder and president of Armstrong/Balmer and Associates, Inc., a position he held until his retirement in 1988. He designed several of Xerox's groundbreaking products, the 914 in 1959, the 813 in 1963, and the LDX, one of the earliest FAX machines. In 1964 he organized Xerox's initial internal design group.

Chris Bangle (b. 1956) U.S automotive designer who was born in Ravenna, Ohio and attended the Art Center College of Design in Pasadena starting in 1977. During summer vacations, he worked for Hartkopf Associates in Wausau, Wisconsin. He graduated from Art Center in 1981 and began his career as designer of interiors at Opel where he became assistant chief designer in 1983. He joined Fiat in Turin in 1984 and by 1992 was head of Fiat design center. He designed the 1993 Fiat Coupe and the 1994 Alpha Romeo 145. In 1992 he was offered a position heading the BMW Design Center in Munich.

Saul Bass (1921-1996). U.S. graphic designer who revolutionized previously mundane opening title credits in films. He studied at the Art Students League in New York and at Brooklyn College under Gyorgy Kepes before working for various advertising agencies beginning in 1946. Important films using his titles were *West Side Story*, *North by Northwest*, *Big* (1994), and *Casino* (1995). In 1952 he established a design firm in Los Angeles, California, Saul Bass & Associates, creating many successful graphic and packaging designs. His firm later became Bass/Yeager & Associates, of which he was chairman and creative director.

Herbert Bayer (1900-1985). Born in Haag, Austria, he studied architecture at Linz and Darmstadt, and painting under Wassily Kandinsky at the Bauhaus from 1921 to 1923. He returned to the Bauhaus as Master of Bauhaus print workshop in 1925, teaching visual communication and typography, and initiated lower-case sans-serif type for all Bauhaus printing. He designed *Universal* typeface in 1927 and left the Bauhaus in 1928 for an advertising agency in Berlin. After organizing a 1938 Bauhaus exhibition at New York Museum of Modern Art, he stayed in the U.S. as consultant to major corporations. His association with the Container Corporation of America (CCA) started in 1946, where he developed the *Great Ideas of Western Man* advertising series and the *World Geo-Graphic Atlas* from 1949 to 1952, a landmark of communication design. He spent his last ten years in California, painting.

Lester Beall (1903-1969) U.S. self-taught graphic designer who was born in Kansas City, Missouri, studied art history at the Lane Technical School and at the University of Chicago, and worked as a free-lance advertising designer in Chicago starting in 1927. He opened an office in New York in 1935 and designed posters for the Rural Electrification Administration, using flat colors and powerful forms which were exhibited at the Museum of Modern Art in 1937. He was art director of Upjohn's magazine *Scope* until 1948. In 1951 he relocated his office to his home in Connecticut.

George A. Beck (1908-1977) U.S. industrial designer who graduated from the Chicago Academy of Fine Arts, worked under Anne Swainson at Montgomery Ward, then at Sears Roebuck in the 1930s. He started at General Electric (GE) in 1934 working under Ray Patten from 1936 to 1942. In 1946 he was named manager of industrial design in GE's new electronics department at Syracuse. In 1957 he established a new industrial design sub-section in GE's Light Military Electronics Department. In 1961, he received ALCOA's annual industrial design award.

Arthur N. BecVar (1911-2003) U.S. industrial designer who graduated with a BFA from Notre Dame in 1933, received an engineering certificate from Purdue University, and an MFA at the Institute of International Education, Charles University, Prague, Czechoslovakia (1933-1934). He worked from 1935 to 1939 with John Gordon Rideout in Cleveland, Ohio and later with Revere Copper and Brass. He joined General Electric in 1944 and headed industrial design there starting in 1948. In 1950, he became manager of industrial design for major appliances, Louisville, Kentucky, where he remained until his retirement in 1976. He was president of ASID in 1955.

Peter Behrens (1868-1940). A German architect and designer, he was born in Hamburg, and studied painting at Karlsruhe and the Düsseldorf schools of art from 1886 to1889. He co-founded the Munich *Vereinigte Werkstätten für Kunst im Handwerk* (United Workshops for Art in Handcraft) in 1897, and was professor of the artists' colony in Darmstadt from 1900 to 1903. He was director of the Düsseldorf School of Arts and Crafts from 1903 to 1907. He was appointed chief artistic advisor at AEG from 1907 to 1914, and was a founder-member of the *Deutsche Werkbund.* Walter Gropius, Le Corbusier, and Mies van der Rohe all studied under him in his office that he opened in 1908. During and after World War I he continued to practice as an architect. In 1922 he was appointed to the Vienna Academy of Graphic Arts and, in 1936, to a Master's architecture program at the Preussische Academy of Art in Berlin. In 1939, he designed a new administration building for AEG in Berlin.

Norman Bel Geddes (1893-1958). Born in Adrian, Michigan, he studied briefly at Cleveland Institute of Art and the Art Institute of Chicago. He began work as an advertising draftsman in Chicago and Detroit in 1913, and established a highly successful career in theatrical set design in New York beginning in the 1920s. He began work as an industrial designer in 1927 at the suggestion of Ray Graham of Graham-Paige Motors Company, and proceeded to design prototype cars for him. He became known as the "P.T. Barnum" of the new field, with futuristic and visionary transportation designs. In 1932, he published his book, *Horizons,* and, in 1939, designed General Motors's *Futurama* at the New York World's Fair. Based on his studies of transportation, he published *Magic Motorways* in 1940, an inspiration for the interstate system of highways.

Mario Bellini (b. 1935) Italian designer who graduated in architecture from Milan Polytechnic in 1959 and opened his own architecture and design office there in 1962. He consulted with Olivetti, Lamy, Artimede and many others. Among his many award-winning designs are the *Divisumma 18* printing calculator of 1972 and the 1973 *Lexikon 82* typewriter, both for Olivetti. Starting in 1986, he became editor-in-chief of *Domus* magazine. In 1987, the Museum of Modern Art exhibited a one-man show of his work accompanied with a Museum publication. A number of his designs are in the Museum's permanent collection.

Lucian Bernhard (1883-1972) Self-taught graphic designer who was born in Vienna, Austria, educated in Switzerland, and in 1903 practiced interior and furniture design in Munich. In Berlin from 1905, he designed posters with rounded, serif letters and product illustrations in a style called *Sachplaket.* Starting in 1913, he designed 36 typefaces, including *Bernhard Roman* (1923) and *Fashion* (1929). In 1920 he became the first professor of poster design at the Berlin Academy of Art. He was involved with establishing the magazine *Das Plakat*, later re-named *Gebrauchsgraphik.* In 1923 he emigrated to the U.S., where he co-founded the design firm *Contempora* with Rockwell Kent (1882-1971), Paul Poiret, and Bruno Paul.

Harry Bertoia (1915-1978) Sculptor and furniture designer who was born in Italy, emigrated to US in 1930, and became a citizen in 1947. He studied at Cranbrook Academy 1937-1940 and became a teacher there. He produced molded plywood products for the war effort for Evans Products Company in Venice, California in 1943, where he also designed chairs with Charles Eames and took a position with him in 1949. In 1950 he established his own studio in Pennsylvania, and became an independent designer in 1952 for Knoll Associates, for whom he designed the *Bertoia Chairs.* After 1953, he devoted himself almost exclusively to sculpture.

Melvin Best (1923-1991) U.S. industrial designer who was born in Pasadena, California and served as a B-17 pilot during World War II. He graduated with a degree in industrial design from the University of Southern California (USC) in 1949, and later received a master's degree in sculpture from California State University. After three years in industry he established his own office in Pasadena, Melvin Best Associates in 1953, and also began teaching at USC. By 1957, his was the largest design office west of Chicago. He was an active member and Fellow of IDSA.

Max Bill (1908-1994) Swiss exhibition, stage, graphic and industrial designer, painter, sculptor, architect, and theorist who studied as a silversmith in Zurich (1924-1927) and in art at the Bauhaus in Dessau (1927-1929). In 1931 he embraced the concepts of Theo van Doesburg and played a major role in the Swiss school of graphic design development. In the early 1950s he designed buildings at Ulm Hochschule für Gestaltung, of which he became director until 1956. He later worked as a freelance designer for watchmakers Junghans and Omega.

Faber Birren (1900-1988) U.S. artist, author, and color consultant who was probably the most influential man of the century in research, education, and the functional use of color in industry, medicine, and the military. He studied at the Art Institute of Chicago and the University of Chicago, settled in Stamford, Connecticut and from the 1920s to the 1980s, lectured and authored over 30 books and 200 articles on all aspects of color and human responses to it. He consulted with House & Garden, Walt Disney Studios, the Hoover Company, and the U.S. Navy, Army, and Coast Guard, among many others.

Robert Blaich (b.1930). U.S. industrial designer who received a BFA in industrial design from Syracuse University. He worked in design and marketing before starting with the Herman Miller Furniture Company in 1953; he served as its vice-president of corporate design and communications from 1968 to 1979. In 1980 he was appointed head of the Concern Industrial Design Centre at Philips Electronics in the Netherlands, where he integrated the disciplines of engineering, marketing, and design, and built a new corporate image of advanced global design. He was a member of IDSA and president of ICSID (1985-1987). In 1990, he received an honorary Doctorate in design from Syracuse University, and, in 1991, was knighted by Queen Beatrix of the Netherlands. He remained at Philips until his retirement in 1992, when he founded Blaich Associates in Aspen, Colorado. He wrote *Product Design and Corporate Strategy* (1993) and *New & Notable Product Design* (1995).

Melvin H. Boldt (1917-1981) U.S. industrial designer who was in partnership with John Morgan from 1947 until 1952 when he left to form Mel Boldt Associates in Chicago, Illinois. He started consulting with Zenith in 1955 upon the death of Robert Budlong. Mel Boldt Associates remained in business until 1990, designing countless consumer products for Zenith, Presto, Brunswick and Moen, among many others.

Eugene Bordinat, Jr. (1920-1987) U.S. automotive designer who graduated from the University of Michigan and studied art at Cranbrook Academy. He joined General Motors in 1939, and became a supervisor of the wartime tank program at Fisher Body. He joined Ford in 1947 as supervisor of advanced styling, working mostly with the Lincoln-Mercury Division. He became vice president of design in 1961 and served in that capacity until his retirement in 1980.

Marianne Brandt (1893-1983) German industrial designer associated with the Bauhaus. She studied painting and sculpture in Weimar, Germany in 1911, became a student at the Bauhaus in 1923, instructed there starting in 1924, and briefly headed the Bauhaus metal workshops in 1928. Her designs as a student were influential to the modern movement. In 1929 she moved to Berlin as a designer in the studio of Walter Gropius. After World War II she taught industrial design at the School of Fine Arts in Dresden (1949-1954), and at an industrial design school in East Berlin (1951-1954) before retiring.

Hin Bredendieck (1904-1995) Industrial designer who was born in Aurich, Germany, enrolled as a Bauhaus student in 1927, and graduated in 1930. He began practice in the Berlin studio of László Moholy-Nagy and Herbert Bayer, then worked at the B.A.G. Lighting Manufacturing Company in Switzerland. Later he returned to Germany to work with Herman Gautel, a furniture manufacturer. In 1937 he was invited by Moholy-Nagy to join the faculty at the New Bauhaus in Chicago and in 1952 became the first director of the new design program at Georgia Institute of Technology, a position he held until his retirement in 1971.

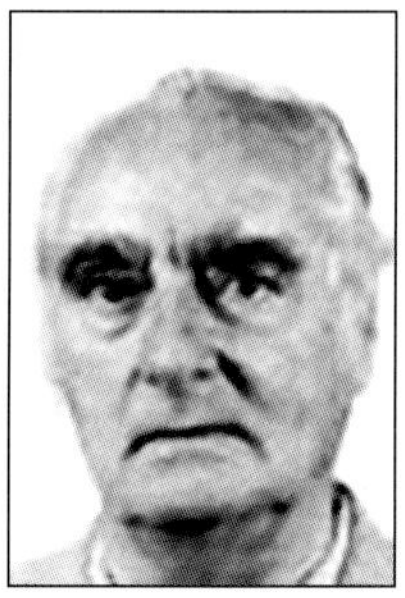

Marcel Breuer (1902-1981). Designer and architect associated with the Bauhaus. Born in Pecs, Hungary, he studied at the Vienna Academy of Art before becoming a student at the Bauhaus in 1920. As a student, he experimented with steel tubing as a construction material for furniture. In 1925 he became Bauhaus master of interior design and worked on a number of furniture designs significant to the modern movement, including chairs of steel tubing. In 1928 he left the Bauhaus to establish his studio in Berlin. In 1937 he was invited by Walter Gropius to the U.S. to teach design at Harvard. He moved to New York in 1946 to establish an office with Herbert Beckhard, Robert Gatje, and Hamilton Smith. Later, he became associated with IBM's design program under Eliot Noyes.

Robert Davol Budlong (1902-1955) U.S. industrial designer born in Denver, Colorado who studied art at Cummings School of Art, Des Moines, Iowa, graduated from Grinnell College, Iowa in 1921, studied art at the Chicago Academy of Fine Art, and opened an advertising office in Chicago. He began his career in industrial design in 1933 with the Hammond Clock Company and consulted with Zenith from 1934 until his death. Budlong's business was then continued by Ken Schory (d. 1991) as Ken Schory Associates. Schory then partnered with Tom Steinbach (Schory-Steinbach Associates, later Creative Corporation), and in the 1970s under Steinbach alone, became the Design Lab.

Gordon M. Buehrig (1904-1990) U.S. automotive designer who studied at Bradley Polytechnic in Peoria, Illinois before becoming an apprentice at the Gotfredson Body Company, Detroit, in 1924. He also worked for Dietrich, Inc., Stutz, and General Motors. He was stylist for the Auburn Automoble Company and in 1936 designed the 810 Cord *Westchester* Sedan and the 812 model that followed, which became design classics. He later worked for Packard, Studebaker, and Ford.

David Chapman (1909-1978). An architectural graduate of Armour Institute of Technology (later Illinois Institute of Technology) in Chicago, he worked on the design staff of the *Century of Progress Exposition* in 1933. That same year, he joined Montgomery Ward as an architect and was appointed Head of Product Design in the Bureau of Design, where he had a staff of eighteen by 1935. He left in 1936 to open his design office, Dave Chapman Inc., later renamed Goldsmith Yamasaki Specht Inc. The firm worked during the war to readapt companies to peacetime production; major clients included International Harvester, Brunswick-Balke-Collender, and Scovill Manufacturing Company. He was president of the Society of Industrial Designers in 1950, and, in 1954, organized a subsidiary company, Design Research, Inc., for long-range planning for clients. In 1955, he returned to Montgomery Ward to re-organize the Bureau of Design.

Serge Chermayeff (1900-1996). Britain's most energetic representative of the Modern Movement, he was instrumental in bringing Modernism to the U.S. Born in Russia, he was educated in London, starting in 1910, as a self-taught architect, designer, and painter. He worked as a journalist (1918-1922), ran a ballroom in Buenos Aires (1922-1924), and was employed by furniture manufacturer Waring and Gillow in London as director of its Modern Art Studio (1924-1927). He entered private practice as an architect from 1931 to 1933, and then as a partner with Erich Mendelsohn from 1933 to 1936. During this period he designed tubular steel chairs for PEL (1931-1934) and the highly successful Ekco radio cabinetsin *Bakelite* (1933). In 1940, he emigrated to the U.S., taught at Brooklyn College (1942-1946), and then replaced Moholy-Nagy as head of Institute of Design, Chicago in 1947. After 1951 he taught architecture at MIT, Harvard, and Yale.

Cesare "Joe Colombo (1930-1971) Italian designer who studied art in late 1940s and architecture at the Politecnico in Milan. He was principally an avante garde painter and sculptor until he set up his own design firm in 1961. He designed products for Candy, Bernini, Stilnovo, O-Luce and Kartell, among others. He won several *Compasso d'Oro* and *Triennale* awards, and in 1972, many of his designs were in the Museum of Modern Art's exhibition, *Italy, the New Domestic Landscape.*

Le Corbusier (1887-1965). Probably the most influential figure in modern architecture between 1920 and 1960, he was born Charles Edouard Jeanneret Gris, the son of a watchcase engraver in La Chaux-de-Fonds in Swiss Jura. In 1907, he met Josef Hoffmann. He worked in the office of Parisian architect Auguste Perret from 1908 to 1909, and with Peter Behrens from 1910 to 1911. He settled in Paris in 1917, took the pseudonym Le Corbusier in 1920, and edited the art journal *Le Esprit Nouveau* from 1920 to 1925, proposing theories of modern design in terms of visual masses and stripping away human experience to reveal pure form, color, and effect. He designed the famous *Le Esprit Nouveau* exhibit at the 1925 Paris *Exposition Internationale*, a range of furniture shown in 1929, and his famous church of Notre Dame-du-Haut at Ronchamp (1950-1954).

Morison S. Cousins (1934-2001) U.S. industrial designer who was born in Brooklyn, NY and graduated from Pratt Institute in 1955. He worked initially for International Harvester, then served in the U.S. Army. In 1963 he formed Morison S. Cousins + Associates with his brother, Michael, in New York. They became known for the European flair of their designs, many of which are in the Museum of Modern Art's permanent collection. The firm became Cousins Design in 1984, and that year, the American Academy in Rome awarded Morison a Rome Prize to study design. In 1990, Morison became Vice President of Design for Tupperware, Inc., a position he held until his death.

Donald Earl Dailey (1914-1997). A U.S. industrial designer, he was born in Minneapolis, Minnesota, and attended art classes at Minneapolis Institute of Art. He studied marketing and engineering at the University of Toledo (1934-1936) and design at the School of Design at the Toledo Museum of Art (1936-1937). He began his career in 1937 as design director for Harold Van Doren and Associates of Toledo, and opened a second office for Van Doren in Philadelphia in 1940. He established his own firm in Philadelphia in 1946, and, in 1950, became product manager at Servel, Inc. In 1952 he was named its vice-president in charge of product planning, and is credited with inventing the term "product planning" and its role for designers in corporations. He left Servel in 1955 to re-establish Don Dailey Associates Inc., in Evansville, Indiana, which he operated until 1995.

Frank Del Guidice (1917-1993) U.S. industrial designer who graduated in advertising design from Pratt Institute in 1938. He spent his entire career with Walter Dorwin Teague Associates starting in 1938. He was sent to Seattle to work on Boeing projects in 1946, was made executive vice president in 1967, and in 1972, became president and chief executive officer. He semi-retired in 1982, remaining as Teague's chairman until 1987.

Donald Deskey (1894-1989). U.S industrial and package designer who was born in Blue Earth, Minnesota, he studied architecture at the University of California, Berkeley, California, and painting at the California School of Design, The School of the Art Institute of Chicago, the Art Students League of New York, and then in Paris (1920-1922). He operated a New York office with Phillip Vollmer from 1927 to 1931, specializing in furniture and lighting. He co-founded the American Union of Decorative Artists and Craftsmen (AUDAC) in 1928. He won the competition for the design of Radio City Music Hall and his work there (1931-1934) popularized Art Deco. He was one of the founders of the Society of Industrial Designers (SID) in 1944. After the war Donald Deskey Associates became known for the bold use of modern materials in interiors and notable packaging for consumer products such as *Joy, Cheer, Gleem*, and *Micrin.*

Freda Diamond (1905-1998). U.S. designer who was born New York City and, after graduation from the Women's Art School at the Cooper Union, taught there while working as a designer in a furniture store. She became a designer and consultant to many U.S. and international manufacturers and department stores in consumer home furnishings. Starting in 1942 she worked for 40 years with Libbey Glass Inc., which by 1954 sold 25 million of products of her design. She wrote *The Story of Glass* in 1953. She designed kitchen canisters for the Continental Can Company, which sold 12 million units, and designed wrought iron furniture for Baumritter, which sold 75,000 pieces in 1954. She worked abroad as an advisor to Italian craftsmen after World War II, to the Japanese government in 1957, to Israel in 1969, and to the World Trade Organization from 1973 to 1981. She was on the advisory board to the Cooper Union Museum, which became the Cooper-Hewitt National Design Museum, Smithsonian Institution.

Raymond Dietrich (1894-1980) U.S. car designer who joined Brewster and Company coachbuilders in 1913, where he studied at the Andrew F. Johnson Technical School in Manhattan and graduated in 1917. In partnership with co-worker Thomas Hibbard, he established *Le Baron Carrossiers* in New York in 1920. He left *LeBaron* in 1925, to become the internal design resource for the Murray Corporation of America. He joined Chrysler as a consultant in 1932, serving as head of its Art and Color Department until his resignation in about 1940. He founded Ray Dietrich Inc. in 1949, and worked with Checker, Lincoln and Mercury until his retirement to Albuquerque in 1969.

Niels Diffrient (b. 1928). U.S. industrial designer, born in Star, Minnesota, who graduated from Cranbrook Academy of Art in 1954. He worked in the office of Eero Saarinen from 1948 to 1953, with the Walter B. Ford office from 1953 to 1954, and with Henry Dreyfuss Associates from 1955 to 1980, becoming an associate in 1967 and a partner in 1970. He was awarded a 1954 Fulbright fellowship to Italy, where he practiced with Marco Zanuso, with whom he won the *Medaglio D'Oro* in 1957 for the Borietti sewing machine. He taught graduate level industrial design at the University of California from 1961 to 1969, and established a studio in Ridgefield, Connecticut, in 1981, concentrating on furniture design.

Jay Doblin (1920-1989). U.S. Industrial designer and educator, who was born in Brooklyn, New York, and graduated from Pratt Institute in 1942. He worked for Raymond Loewy (1942-1955) directing the Frigidaire account and designing vending machines for Coca-Cola, razors for Schick, and pens for Eversharp. He was briefly a partner with Lippincott & Margulies and directed the night school at Pratt from 1947 to 1952. He became director of IIT's Institute of Design in Chicago from 1955 until 1969. He published several books including *Perspectives, A New System for Designers* (1955) and *One Hundred Great Product Designs* (1970). He was president of ASID in 1956, and of IDEA in 1962. In 1969 he joined Unimark, a firm specializing in corporate identity programs, as senior vice-president until 1972, when he left Unimark and founded Jay Doblin & Associates.

Donald R. Dohner (1897-1943). U.S. industrial designer who grew up in Indiana and studied at John Herron Art Institute, Indianapolis, the Academy of Fine Arts, Chicago, and the Art Institute of Chicago. In 1918 he went to Pittsburgh to study set design at Carnegie Institute of Technology (CIT) (now Carnegie Mellon University), but while waiting for classes to begin, took a teaching job as industrial arts instructor in the Pittsburgh public school system, then became a faculty member of CIT. He started with Westinghouse as a consultant in 1926 as an "Art Engineer," and was hired as director of art in 1929. In 1934, he initiated the world's first degreed program in industrial design at CIT. He left CIT in 1935 to initiate a similar program at Pratt Institute in Brooklyn, becoming its supervisor. In 1943 he opened a design office with J. Gordon Lippincott, Dohner, and Lippincott. Dohner died on Christmas Eve, 1943, and was replaced at Pratt by Alexander Kostellow.

Henry Dreyfuss (1904-1972) U.S. industrial designer, was born in Brooklyn, New York, to a family in the theatrical materials supply business. He completed studies as an apprentice to Norman Bel Geddes in 1924, and produced 250 theatre stage sets for a number of theatres between 1924 and 1928. He opened his office in 1929 for stage and industrial design activities. He focused on design problems related to the human figure, working on problems from "the inside out," and believed that machines adapted to people would be the most efficient. Major clients included Bell Telephone Labs, AT&T, John Deere, and Polaroid. Several of his books were published, including an autobiography, *Designing for People,* in 1955 that included his famous "Joe" and Josephine" anthropological charts, and *The Measure of Man* in 1960. He became the first president of IDSA upon its formation in 1965. Retired in 1969, he died in Pasadena, California.

Charles Eames (1907-1978). U.S industrial and graphic designer who was born in St. Louis, Missouri, and studied architecture at Washington University (1924-1926). He worked with an architectural firm in St. Louis (1925-1927), and then as a partner of Gray and Eames (1930 -1937). In 1937 he became head of the department of experimental design at Cranbrook Academy of Art and worked with Eero Saarinen investigating plastics. They won a MoMA furniture design competition in 1941. He married Ray Kaiser in 1941 and moved to Los Angeles to design film sets for MGM, but continued his experiments with plywood. Their workshop became part of the Evans Products Company in 1943, laminating glider parts. At war's end, the company turned to production of Eames's chairs. He sold rights for his chairs to the Herman Miller Furniture Company in 1946. Eames produced some of the most creative design work of the century in furniture, film, and exhibits.

Ray Kaiser Eames (1916-1988) U.S. graphic and industrial designer who trained as a painter with Hans Hofmann in the 1930s, had her work shown in 1936, and was a founding member of the American Abstract Artists. She studied at Cranbrook Academy 1940-1941, married Charles Eames, and practiced as a graphic designer. Between 1942 and 1944, she designed more than half the covers for the new magazine *California Arts and Architecture.* After World War II, she became her husband's professional creative partner in a range of design projects, and in 1981, was awarded honorary membership in the Industrial Designers Society of America.

Harley J. Earl (1893-1969). U.S. automotive designer, was born in Hollywood, California, attended the University of Southern California in 1912, and studied engineering at Stanford University from 1914 to 1917. He began his career with the firm his father founded in 1908, the Earl Automobile Works, that built custom car bodies for film stars starting in 1911. Don Lee, West Coast Cadillac distributor, bought the company and employed Harley in 1919. Larry Fisher, Cadillac's division president, soon recognized Earl's talents and he was sent to Detroit in 1925. He established General Motors's Art and Color Department in 1927, became vice-president in 1940, and dominated GM design policies until his retirement in 1958. In 1945 he established his own firm, Harley Earl Inc., to design products in non-automotive fields, which in 1964 merged with Walter B. Ford Design Associates, Inc., to form Ford & Earl Design Associates.

Philip S. Egan (b.1920) U.S. industrial designer who studied aeronautical engineering at Stewart Technical Institute, New York and started work with Harry Preble, Jr. in 1941. He served in the U.S. Army Air Corps from 1943 until 1946, when he joined J. Gordon Lippincott and was assigned to the *Tucker '48* project, working with Alex Tremulis at the Tucker Corporation. Starting in 1948 he worked with Sears, Roebuck & Company on a variety of products and in 1960, opened his office in Fairfax, California, Phil Egan Design. His book, *Design and Destiny: The Making of the Tucker Automobile,* was published in 1989.

Fritz Eichler (1911-1991) German industrial designer and executive who began his career in theatre design. Engaged by Artur and Irwin Braun in 1955, he brought Braun AG to the forefront of good design through his selection of internal staff, and through links between Braun and the Hochscule für Gestaltung at Ulm.

Elwood P. Engel (d. 1986) U.S. automotive designer who graduated from Pratt Institute and joined General Motors as an engineer in 1940. After the war, he worked for George Walker and joined Ford with Walker in 1955 when Walker became Ford's vice president of styling. In 1961 he joined Chrysler, succeeding Virgil Exner as head of design, where he remained until his retirement in 1973, and after that, as consultant until 1974.

Hartmut Esslinger (b. 1945) German industrial designer who was trained in engineering and design and established his office in Altensteig, Germany in 1969. He named it *frogdesign* in 1982, when he opened additional offices in California and later in Tokyo. His design of the *Apple IIc* computer and the *NeXT* workstation for Steve Jobs catapulted *frogdesign* into international recognition, and Esslinger onto the cover of *Business Week* magazine in 1990, headlined as *Rebel with a Cause.*

Virgil Max Exner (1909-1973) U.S. automotive designer who was born in Ann Arbor, Michigan and studied art at Notre Dame University. Hired by Harley Earl at General Motors in 1933, he became Pontiac styling chief and designed its "Silver Streak" hood ornament. In 1937 he joined Raymond Loewy Associates, and in 1939 headed the Studebaker account in South Bend. In 1944, he was fired by Loewy and hired directly by Studebaker. In 1948 he returned to Detroit to work for Chrysler, was named director of styling in 1953, and developed their 1957 *Forward Look.* He became vice president of Chrysler in 1957 and served in that capacity until 1961.

Montgomery Ferar (1910-1982) U.S. industrial designer who was born in Boston, received a bachelor's and master's degrees in architecture at MIT in 1932, and studied in Europe in 1933. He worked for several architectural firms in Boston before moving to Detroit to join General Motors (GM) Art and Color Section where he met Carl Sundberg. In 1934 they both left GM to form Sundberg-Ferar, Inc., designing for Sears,Roebuck and the Whirlpool Corporation. In the 1970s, they designed mass transit systems for many U.S. cities. They both retired in 1975, but in 2000, Sundberg-Ferar Inc. still maintained four U.S. offices and a location in the Far East.

Gordon William Florian (1909-1984) U.S. industrial designer who was born in Connecticut and studied at Yale University School of Fine art in New Haven, Connecticut, and Grand Central School of Art in New York. He left school in 1931 to work for General Electric (GE), where he worked in marketing. In 1935 he joined GE's first Appearance Design Department, becoming assistant director in 1940. In 1946 he opened private practice in Bridgeport, Connecticut and in 1949, with others, he established a four-year industrial design degree program at the University of Bridgeport. From 1950 to 1954 he lectured to engineering and business students at MIT.

Paul Theodore Frankl (1887-1958) Austrian furniture designer and member of the *Secession* movement of 1897 who emigrated to the U.S. in 1914 and established a gallery selling his own work in 1922. He established offices in New York and Los Angeles and was known for his "skyscraper" style furniture in the late 1920s. He wrote *New Dimensions* in 1928, and *Form and Re-form* in 1930. He designed pieces for the Johnson Furniture Company in the 1940s and after the war relocated to California, working in a style typical of "California Modern."

Richard Buckminster Fuller (1895-1983). One of the century's most intellectual and radical inventor/designers, he studied mathematics at Harvard 1913-1915, worked with Armour & Co. until 1922, when he became president of a wall-building systems company. In 1927, he formed the 4-D Company to develop his inventions, including his *Dymaxion Car*, prototyped in 1933, his *Dymaxion* Bathroom, prototyped in 1938, and his *Dymaxion House*, produced as the *Wichita House* in 1946. In 1949, he started development of his geodesic dome, patented in 1954, which was used extensively for exhibition enclosures, including the U.S. pavilion at Expo '67 in Toronto. From 1959, he was research professor at Southern Illinois University. In 1969, he published *Operating Manual for Spaceship Earth* and was nominated for the Nobel Peace Prize. In 1985, he was posthumously honored with the naming of a newly created form of carbon molecule (C 60) as *buckminsterfullerene* (nicknamed *buckyballs*), and others in its class (*fullerenes*), such as *buckeytubes*, discovered in 1991.

James Franklin Fulton (1930-2003). U.S. industrial designer who was a 1951 graduate of Pratt Institute. He worked for the Towne Silver Company in 1951, and the same year joined Owens-Corning Fiberglas Corporation in Toledo as its first staff designer. He worked from 1953 to 1958 as senior designer with the office of Harley Earl, Inc. In 1958 he was appointed design director with Raymond Loewy's Paris office, and in 1960 became director of product design and transportation with Loewy/Snaith Inc. in New York, becoming vice-president in 1962 and senior vice-president in 1964. He established his office in New York, Fulton + Partners, Inc., in 1966, and added a Paris affiliate, Endt + Fulton in 1975. He was president of IDSA, 1975-1976 and was chairman of Design Publications, Inc., which he co-founded in 1977 to publish ID magazine. In 1969 he became a trustee of Pratt Institute, and served as its chairman from 1991-2001.

Frank O. Gehry (b. 1929) Canadian-born US architect and designer who was trained at the University of Southern California and at the Harvard Graduate School of Design. He worked for several West Coast architectural offices before establishing his own practice in 1962, pioneering in unusual broken forms and angles. He is generally associated with *Deconstructivism* and *Post-Modern* movements, but has also been active in furniture and product design. Including his1972 *Easy Edges* furniture of corrugated cardboard and 1983 Gehry lamps featuring fish forms.

Tom Geismar (b. 1931) U.S. graphic designer who concurrently attended Brown University and Rhode Island School of Design, graduating in 1953, and then received a master's degree in graphic design from Yale University School of Art and Architecture in 1955. With Ivan Chermayeff, he founded Chermayeff & Geismar, Inc. in 1960. He chaired the U.S. Department of Transportation's advisory committee that developed a national system of standardized symbols 1974-1976. He was a professor at Cooper Union in New York and in 1998 authored and designed, *Spiritually Moving*, a book on American folk sculpture. In 2000, Princeton University Press published an anthology of over 100 trademark designs by Chermayeff & Geismar.

Francesco Gianninoto (1903-1988) U.S. package designer who began his career as a commercial artist c. 1920 without formal training, becoming art director at BBD&O. He established Gianninoto Associates in 1932, of which he was president until 1983. He was a founder of the Package Design Council and a member of IDSA. He was an honorary professor at Western Connecticut State University where he taught communication through design. Two of his most famous designs were the Marlboro cigarette package (mid-1950s), and "Elsie Daisy" the Borden Cow. He retired in 1983 to Fairfield County, Connecticut.

Alexander Girard (1907-1993). U.S. architect and interior designer who trained in London, Florence, and Rome, returning to the U.S. in 1932 to open a design office in New York. In 1937 he relocated to Detroit and continued to bring color and pattern to interiors and products during the 1950s and 1960s that were counter to the International Style. Starting in 1952 he was design director of upholstery for the Herman Miller Furniture Company. He designed Braniff's creative brand identity program as well as livery for its planes. His extensive collection of folk art was donated to the Museum of International Folk Art, Santa Fe, New Mexico, in 1976, for which he designed a new wing and permanent exhibition.

Giorgetto Giugiaro (b. 1938) Influential Italian car designer and key figure in the firm of ItalDesign whose grandfather and father were painters of frescos. He studied at Turin Academy of Fine Arts before joining Fiat as an apprentice in 1965 at age 17, where his work included the Fiat *Spider* of 1965. In 1968 he became head of styling at Carrozzeria Bertone. He worked after that briefly at Ghia as lead designer and in 1968 founded ItalDesign with former Fiat engineer Aldo Mantovani. In 1981 he founded ItalDesign-Giugiaro which expanded into consumer products, and in 1987 founded Guigiaro SpA for clothing and accessories.

Henry P. Glass (1911-2003). U.S. industrial designer who was born in Vienna, Austria, and studied architecture at the Technical University of Vienna, 1929-1936. He emigrated to the U.S. in 1939 and worked for Gilbert Rohde and Russel Wright. In 1940 he moved to Chicago and studied under László Moholy-Nagy and Gyorgy Kepes at the Institute of Design. In 1946 he opened a studio in the Furniture Mart and began teaching at the School of the Art Institute of Chicago, where he served as a full professor until 1969. He joined the American Designers Institute in 1939, was chair of the Chicago Chapter of ASID, 1959-1960, and became a Fellow of IDSA in 1965.

William M. Goldsmith (1917-1987) U.S. industrial designer who was born in Rochester, New York and graduated from Carnegie Institute of Technology in 1939. He joined the Chicago firm of Dave Chapman, Inc., becoming a partner and eventual chairman of the board as the firm became Chapman Goldsmith Yamasaki Inc. in 1955, then Goldsmith Yamasaki Specht, Inc. in 1970. He was president of IDSA 1971-1972, an IDSA Fellow, and received the Society's Personal Recognition Award in 1980.

Clarence F. (Cal) Graser (1923- 2000) U.S. industrial designer who was born in Toledo, Ohio, served with the Army Air Force in World War II, and graduated from the University of Michigan in 1948. He worked as an industrial designer for General Electric, then Joined IBM in 1959 as manager of industrial design. In 1964 he joined Westinghouse as director of a new Consumer Products Design Center in Columbus, Ohio, until Westinghouse sold their consumer products business in 1974. He then joined Richardson/Smith in Ohio until his retirement to Venice, Florida in 1978.

Eileen Gray (1878-1976) Irish interior, furniture, and architectural designer who studied at the Slade School of Art in London before moving to Paris in 1902, where she became expert in Japanese lacquer techniques. In 1909 she crossed the English Channel in a balloon with Charles S. Rolls of Rolls-Royce fame. After World War I, she expanded her work and opened a gallery in Paris in 1922. In the 1930s she designed in *Art Deco* style with chromium-plated metals and glass. She was forgotten after World War II but re-discovered by the Museum of Modern Art, which mounted an exhibition of her work and produced a concurrent book in 1980.

Eugene Turrenne "Bob" Gregorie, Jr. (b. 1908) U.S. automotive designer who was born in New York City in 1927, started as a draftsman at Elco Boat Works in Bayonne, New Jersey, then worked with yacht designers Cox & Stevens in New York in 1928. In 1929 he joined General Motors but lost his job in the Depression. In 1931 he joined the Lincoln Motor Company as a draftsman and in 1935 was placed in charge of the new styling group at Ford. He remained until 1946 when he retired to sail on a trawler he designed. In the mid-1970s he retired to St. Augustine, Florida.

Walter Adolph Gropius (1883-1969). German architect and educator with a pivotal role in the modern movement, who was born in Berlin, studied architecture in Munich and Berlin, and worked in the office of Peter Behrens. He joined the Deutsche Werkbund in 1910, and supported Henry van de Velde on the principle of the creative individual over standardization. After serving in World War I, he became director of the Weimar School of Arts and Crafts where he established the Bauhaus, considered by many to be the first educational program of modern design. He resigned from the Bauhaus in 1928 and set up an office in Berlin. After working in London, he came to the U.S. in 1937 to become a professor at the graduate school of design at Harvard, working in partnership with Marcel Breuer. In 1945 he founded The Architects Collaborative (TAC). Gropius received the Kaufmann International Design Award in 1962 for achievement in design education.

Robert Gruen (1913-1999) U.S. industrial and package designer who received an illustration degree from Carnegie Institute of Technology in 1934, worked in 1935 as a set designer for Norman Bel Geddes, and for a film studio in Hollywood in 1936. He returned to New York in 1937 to work as a designer of glassware and opened his own office in 1940. A founding member of the Package Designers Council in 1952, he became president of IDI in 1953-1954, and was instrumental in its merger with ASID to form IDSA in 1965. He retired to Philadelphia, Pennsylvania.

Lurelle Guild (1898-1985) U.S. industrial designer who received a degree in painting in 1920 from Syracuse University and began his career in the furniture and antiques field. He soon began illustrating and writing for magazines. He opened his own office in 1924 and in 1928 titled his business "Guild of Industrial Design". His most famous design was the 1937 Electrolux cleaner. He was one of the founders of the Society of Industrial Designers in 1944.

Olle E. Haggstrom U.S. industrial designer who received a certificate of industrial design from Pratt Institute, and a B.S. in industrial design from the University of Bridgeport. He worked in major appliance design at General Electric (GE) from 1946 to 1960 and in 1960 was named manager of General Electric Housewares industrial design, where he remained until his retirement in 1985.

Tom Hardy U.S. industrial designer who graduated in industrial design from Auburn University, joined IBM in 1970, and became corporate manager of the IBM design program, directing worldwide identity activities and coordinating fifteen design centers around the world. He then became an independent design strategist for Samsung and Polaroid.

Marc Harrison (1936-1998) U.S. industrial designer who was a long-time faculty member and chair of the industrial design department at the Rhode Island School of Design (RISD). In addition, he maintained an active design consultancy, focusing on the needs of the physically challenged long before the term "Universal Design" was in use. He was the designer of the Cuisinart food processor in 1979. In 1997, RISD honored his contributions to the school during the dedication of its new Industrial design building, and the RISD Alumni Association named him 1997 Educator of the Year. In 1998, Marc succumbed to his two-year battle with ALS.

Jon William Hauser (1916-1999) U.S. industrial designer who was born in Sault St. Marie, Michigan, and began his career with General Motors in 1936. He worked for Chrysler (1941-1943), Sears, Roebuck as director of design (1943-1945), Dave Chapman (1945-1946), Barnes & Reinecke (1946-1949), Reinecke Associates (1949-1952), and as executive vice president of Raymond Loewy's Chicago office, (1960-1961). He established his Chicago office, Jon W. Hauser, Inc, in 1952 which became Hauser Design, Inc. in 1970, and was headed later by his son, Jon W. Hauser, II. He was president of IDI 1962-1963. After retirement, he served as visiting professor of design at the University of Illinois in Champaign-Urbana.

Samuel M. Highberger U.S. industrial designer who graduated from Carnegie Institute of Technology (now Carnegie Mellon University) in 1949. From 1950 to 1959 he worked for Sundberg & Ferar, Inc. and from 1959 to 1985, was product design director of Ford & Earl Design Associates. Many of his designs won *Design in Steel* awards (1968-1973) and he was the youngest designer to win an Industrial Designers Institute (IDI) national award in 1958. In 1965, he established his consulting firm, Highberger Design, in Birmingham, Michigan, which in 1985 became Hi Design, Inc. until his retirement to Hendersonville, North Carolina.

Gerald P. Hirshberg. (b. 1940) U.S. automotive designer who earned degrees in mechanical engineering at Ohio State University, and in industrial design at the Cleveland Institute of Art. In 1959 he had a hit rock & roll song, *Sparkling Blue*, but gave up music for automotive design. He joined General Motors in 1963, where he worked on the 1967 Pontiac *Firebird*. In 1980 he was named director of design at Nissan Design International in La Jolla, California, the first of many West Coast car design studios, and in 1985 was named its president. He retired in 2000 and wrote the book, *Creative Priority*.

Josef Hoffmann (1870-1956). Austrian architect and former student of Otto Wagner, who studied architecture in Munich and Vienna, and worked with Wagner from 1895 to 1899. He was a professor at the Viennese School of Fine Arts (1899-1941) and co-founder in 1903 of the *Wiener Werkstätte*, for which he produced designs until 1931. He designed several chairs for Thonet, but his major architectural masterpiece was the Palais Stoclet in Brussels (1905-1911).

Robert H. Hose (1915-1977). U.S. industrial designer who was born in Sleepy Eye, Minnesota, graduated from the University of Minnesota in 1937, and completed his Masters in architecture at MIT in 1940. He started at Bell Labs in 1939, working under industrial design consultant Henry Dreyfuss, and joined the Dreyfuss organization in 1946, later becoming a partner in the firm. He contributed to the design of a number of telephones, including the 500 (1949), the wall mounted phone (1956), the first push-button phone (1958), and the *Princess* phone (1959). He was president of the Society of Industrial Design (SID) in 1953. He left Dreyfuss in 1961 to open his own consulting firm. A Fellow of SID, he worked with other design organizations in planning the merger that formed the Industrial Designers Society of America (IDSA) in 1965, and served as president of the new organization from 1967 to 1968.

Elbert Hubbard (1856-1915). U.S. businessman and writer who was born in Bloomington, Illinois. He founded the very successful Larkin Soap Company in 1875. In 1883 he sold his interest in the company and published *The Philistine,* a successful magazine in East Aurora, New York. He was inspired by William Morris and the Arts and Crafts Movement in England, founded the Roycroft workshops in 1896 to produce high quality furniture and, in 1897, a print shop to promote his craft philosophy, "Not How Cheap But How Good." A dashing figure with long hair, broad-brimmed hat, cane, and loose black satin bow tie, Hubbard's unconventional costume would be emulated by Frank Lloyd Wright during Wright's entire career. Hubbard and his wife perished on the *Lusitania,* but his son Elbert II maintained Roycroft until 1938.

Leonard Garth Huxtable (1911-1989) U.S. industrial designer who was born in Nova Scotia and graduated from the Massachusetts School of Art in 1933, and worked for Norman Bel Geddes and Egmont Arens in New York. In 1939 he worked for Albert Kahn and General Motors in Detroit, then in 1941 for Benjamin Webster and Henry Dreyfuss in New York. He opened his own office in 1948 and designed many award-winning hand and power tools over a 20-year period for the Millers Falls Company. He opened a joint office in 1952 with his wife Ada Louise, an architectural critic for the New York Times.

Alfonso Iannelli (1888-1965) Sculptor, graphic artist, interior and industrial designer who was born in Italy, came to the U.S. in 1898 and trained with sculptor Gutzon Borglum in New York. He went to Chicago in 1914 to work with Frank Lloyd Wright on Midway Gardens and founded Iannelli Studios there in 1915. Beginning in 1933, he entered the new field of industrial design and from 1948 to 1954 designed a range of products for the John Oster Manufacturing Company, including mixers, hair dryers, knife sharpeners, and hair clippers.

Arne Jacobsen (1902-1971). Danish architect who studied at the Royal Danish Academy of Fine Arts (1924-1928). He worked in Copenhagen as a city planner, architect, and designer. He lived in exile in Sweden during World War II, but began the design of products in 1951 when he became interested in Eames's experiments in plywood. He gained international attention with his *Ant* side chair (1952), *Swan* chair (1957), and *Egg* chair (1957). From 1956 to 1971 he was professor of architecture at the Academy in Copenhagen.

Paul Jaray (1889-1974) Austro-Hungarian engineer who started working in 1914 at the Flugzeugbau Friedrichshafen, Germany, a seaplane off-shoot of the Zeppelin Works. In 1916 he built a wind tunnel for aircraft testing, and in 1921 began testing automobile models. He built a full-scale model and filed patents that were granted in 1927 for a highly streamlined car. A production car, the Tatra 77, was built in 1933 using the Jaray design, but many manufacturers simply out-waited patents rather than paying Jaray for design rights.

Georg Arthur Jensen (1866-1935). The best-known Danish silversmith, he studied sculpture at Copenhagen Academy and, beginning in 1895, worked in ceramics with Joachim Petersen, where the Arts and Crafts movement heavily influenced him. He opened a shop in 1904 to produce and retail his work; he was joined by Harald Nielsen and Johan Rohde (1856-1935). The Jensen firm opened a U.S. branch in 1922 in New York. Count Sigvard Bernadotte of Sweden later joined the firm. Jensen's son, sculptor Søren George Jensen (b. 1917), headed the firm from 1952 until 1974, when he retired.

Ivar Jepson (1903-1968) Swedish born inventor and engineer who studied engineering at Heslehom Technical High School and took graduate work at the University of Berlin. He emigrated to the U.S. and by 1925 was employed as a draftsman by the Chicago Flexible Shaft Company, which later became the Sunbeam Corporation. From 1930 on he produced a great number of appliance patents, became chief engineer in 1932, and vice president of Research & Development in 1956, a position he held until his retirement in 1963.

Albert Kahn (1869-1942). U.S. architect, was born in Germany, but came to U.S. as a child. He worked for a Detroit architectural firm before opening his own office in 1902. He pioneered in industrial architecture in concrete and steel, and set the prototype of 20th century factory architecture. His designs were modernist, characterized by lack of ornamentation and large areas of glass. He designed buildings for Ford and Chrysler, and designed grain elevators for Kellogg in Battle Creek, Michigan, which were greatly admired by modernists.

Wassily Kandinsky (1866-1944). Along with Kasimir Malevich and Piet Mondrian, he was one of the three founding fathers of 20th century abstract painting. He was born in Moscow and in 1896 broke off a career in science to take up painting. He studied in Munich and was a member of the Berlin Secession movement. He founded *Neue Künstlervereinigung* in 1909 and the *Blaue Reiter* (with Franz Mark) in 1911. He returned to Moscow during World War I, as his sympathies were with the Revolution, and was made a professor at the re-organized Academy of Fine Arts in Moscow in 1918. He became a member of Commissariat for Education and director of the new State Museum of Pictorial Culture in 1919. In 1920 he was appointed professor at the University of Moscow and in 1921 founded the Academy of Arts and Sciences, becoming its vice-president. He returned to Germany in 1921 and was called to teach at the Bauhaus in 1922. He remained there until 1933, when he emigrated to France.

Edgar Kaufmann Jr. (1910-1989). U.S. industrial design spokesman and historian who was the son of Edgar Kaufmann Sr., a wealthy Pittsburgh merchant who founded a department store there. He attended Kunstgewerbeschule of the Oesterriechisches Museum für Angewandte Kunst in Vienna in the late 1920s, studied painting and typography with Victor Hammer in Florence, and was an apprentice architect at Frank Lloyd Wright's Taliesin Foundation (1933-1934). He strongly supported his father's decision to commission Wright to design *Fallingwater,* the Kaufmann summer home. In 1940 he wrote to Alfred Barr of the Museum of Modern Art (MoMA), proposing an *Organic Design in Home Furnishings Competition.* He served in the U.S. Air Force, and joined the MoMA in 1946 as director of the industrial design department, a position he held until 1955. He initiated MoMA's *Good Design* program in 1950.

Henry Keck (b. 1921) U.S. industrial designer born in New York City who graduated with majors in engineering and business from Dartmouth and entered military service, studying advanced electronics. In 1947 he received a degree in industrial design and engineering at California Institute of Technology in Pasadena, California. After a brief stint at Corning Glass Works, he became assistant manager of the West Coast Office of Raymond Loewy. In 1951 he formed his own firm of Henry Keck Associates in Pasadena and was joined by a partner, mechanical engineer Burnie Craig. In 1953, the firm was re-named Keck-Craig, Inc.

David M. Kelley, U.S. industrial designer who received a degree in Engineering from Carnegie Mellon University, and a graduate degree in product design from Stanford University. He founded Hovey-Kelley Design, which worked on improvements to the Engelbart computer mouse for Apple. In 1984 he established Dave Kelley Design in Palo Alto, California, and Onset, a partnership to capitalize design entrepreneurs. He teaches engineering at Stanford University and in 1990, with Bill Moggridge, became a co-founder of IDEO, the highly successful design firm.

Gyorgy Kepes (b. 1906) Hungarian-born painter, designer, educator and author who graduated from the Royal Academy of Fine Arts, Budapest, in 1928. He worked in stage and exhibit design in London and in the Berlin studio of Laszlo-Moholy-Nagy (1936-1937) before coming to the U.S. in 1937. He taught at the Institute of Design in Chicago (1937-1946) and from 1946 until his retirement in 1974, was professor of visual design at MIT. He wrote *Language of Vision* in 1944, and was guest editor of the *Vision+ Value* Series by MIT Press in 1966. His paintings are in many major museums.

Walter Maria Kersting (1892-1970) German architect, designer, graphic artist and engineer who attended Technische Hochschule in Hanover 1912-1914 where he worked as a graphic artist. He worked 1922-1932 as an architect in Weimar and Cologne, and designed furniture, lighting and radios, in particular the 1933 *Volksempfänger* (People's Radio). He wrote *The Living Form* in 1932 and taught graphics and book design 1933-1944 at Staatliche Kunstakademie in Düsseldorf He then opened an industrial design studio and designed a variety of products including audio equipment, sewing machines and telephones into the 1950s.

Florence Knoll Bassett (b. 1917). U.S. architect and designer, who was born Florence Schust in Saginaw, Michigan. She studied at Cranbrook Academy of Art, the Architectural Association in London, and at the Illinois Institute of Technology under Mies van der Rohe. She worked briefly in Chicago before going to New York, where she met and married Hans Knoll, co-founding Knoll Associates in 1946. She established the Knoll Planning Unit, an interior design service of Knoll. Through her, Knoll Associates had contacts with leading figures in architecture and design. She played an important role in selling modern furniture design to postwar corporate America. She served as president of Knoll Associates from 1955 until 1959, when she sold her interests, but remained as design director until her retirement in 1965. She married H. Hood Bassett in 1958 and divides her time between Florida and Vermont. In 2002, she earned the Medal of Arts for her dedication to 20th century modern design.

Hans Knoll (1914-1955) U.S. businessman who was born in Stuttgart, Germany where his father was a furniture maker. He studied in Switzerland and England and came to the U.S. in 1937 to establish his own business, the H. G. Knoll Furniture Company, which offered simple, modern furniture. He married Florence Schust in 1944 and formed Knoll Associates in 1946.

Belle Kogan (1902-2000). Believed to be first female industrial designer in the U.S., she was born in Ilyashevka, Russia, and emigrated to the U.S. in 1906. In 1920 she began working in her father's jewelry store while studying at Pratt Institute and the Art Students League. In 1929 she was employed by the Quaker Silver Company, which trained her as a silver designer at New York University, Rhode Island School of Design, and at a Pforzheim, Germany, Art School. She soon headed a department of twenty at Quaker in Attleboro, Massachusetts. She opened her own New York office, Belle Kogan Associates, in 1932, and designed many houseware products. In 1944 she was Secretary Treasurer of IDI. She closed her New York office in 1970 and moved to Israel under contract with KV Design, setting up a studio offering comprehensive design services. In 1972 she left the studio to work as a consultant.

Alexander Jusserand Kostellow (1896-1954). A design educator born in Persia (Iran), he studied in Paris and at the University of Berlin. He came to U.S. in 1916, where he studied painting at the National Academy, at the Art Students League, and at the Kansas City Art Institute, where he met and married Rowena Reed. Both came to Pittsburgh in 1929 to teach at Carnegie Institute of Technology, he in painting and she in sculpture. In 1934 he helped Donald Dohner initiate an industrial design program at CIT. He left in 1936 for Pratt Institute where he headed the program in 1944, after Dohner's death, and remained there until his own death.

Rowena Reed Kostellow (1900-1988) U.S. sculptor, designer and educator who was born in Kansas City, Missouri and received a journalism degree from the University of Missouri. She studied sculpture at the Kansas City Art Institute, and painting at the Art Students League. She married Alexander Kostellow and in the late 1920s taught with him at Carnegie Institute of Technology in Pittsburgh, Pennsylvania. In 1938 with her husband at Pratt Institute, she taught industrial design, chaired the design department from 1962-1966, and continued teaching until 1987.

Homer LaGassey (b. 1934) U.S. automotive designer who received the first General Motors (GM) Industrial Design Scholarship, and started at GM in 1942. After serving as a World War II bomber pilot, he graduated from Pratt Institute in 1947 and returned to GM. He joined Chrysler in about 1954, and Ford about 1959, where he managed the *Mustang* studio 1968-1972. In 1959 he established an automobile design program for the Center for Creative Studies in Detroit, where he taught for 28 years. He retired to paint World War II air battles, and in 1994 was inducted into the Automotive Hall of Fame.

Thomas Babbit Lamb (1897-1988) U.S. inventor, industrial designer and cartoonist who was born in New York City, studied at the Art Students League and New York University while working for a textile studio, and served in the U.S. Navy in World War I. In the 1920s, he wrote and illustrated children's books and had a monthly page in *Good Housekeeping* magazine called "Kiddyland Movies", which spawned a line of children's products. He became interested in hands through drawing and the study of anatomy. Starting in 1941 he developed ergonomic handle designs that brought him design acclaim. He died in New Canaan, Connecticut.

Harper Landell (1918-1997) U.S. industrial designer who attended Philadelphia Museum School and taught at Rutgers University for two years. He started his career in 1939 with General Motors and worked in their aircraft division on warplanes during the war. In 1947 he was hired by Harold Van Doren and Associates in Philadelphia and upon Van Doren's death in 1957 he continued the office as Harper Landell & Associates. In 1975 Woodstream Corporation bought his office and he moved into the company as designer and head of research until 1987. He retired to Florida in 1990.

Walter Landor (1913-1995). A graphic designer born in Munich, Germany, he was founder-partner of the Industrial Design Partnership in London (1935-1939). Traveling to the U.S. in 1939, he became associate professor of industrial design and interior architecture at the California College of Arts and Crafts in 1940. In 1941 he founded Walter Landor Associates in San Francisco, which would later establish regional headquarters in New York, London, and Tokyo, specializing in corporate identity, branding, and packaging. In 1964 he moved to a floating office, the *Klamath* ferryboat at San Francisco Pier 5, which author Tom Wolfe dubbed, "the flagship of packaging design."

Richard S. Latham (1920-1991). U.S industrial designer born in Kansas City, Missouri, who studied engineering at Kansas City Engineering School and design at Armour Institute in Chicago under Mies van der Rohe, 1940-1942. He started with Montgomery Ward in 1942, and, after his military service, joined Raymond Loewy's Chicago office in 1945, working on Greyhound buses and aircraft interiors. He later became director of design for Loewy's Chicago office. He spent five years with the German porcelain industry, establishing Rosenthal's *Studio Line* in 1954. In 1955 he founded Latham Tyler Jensen, Inc. with two other Loewy designers, Robert D. Tyler and George Jensen. He was president of ASID in 1959 and of ICSID in 1965. In about 1970 he founded Richard S. Latham & Associates, Inc., where he pioneered in product planning, and became design advisor for Bang & Olufsen of Denmark and for Land's End, which he helped found.

George Lawson U.S. automotive designer who graduated from the Cleveland Institute of Art and started in commercial art. He worked at Buick in the late 1930s, where he designed the 1938 Buick. He then worked at Briggs Manufacturing Company briefly and during the war, returned to General Motors to design and produce training manuals for government projects. He met Preston Tucker in 1946 and prepared the initial Tucker *Torpedo* design concept. Later he worked for Raymond Loewy Associates, Nash, and General Motors in the 1950s.

Michael Lax (1929-1999) U.S. industrial designer who graduated from the New York High School of Music and Art in 1947 and received a degree in design from Alfred University in 1951. He worked for Dunbar Glass Company until 1954, then, after a two-year Fulbright Fellowship to Finland, he returned as a consultant to the Raymor Company in 1956. He designed products for Copco for 35 years, starting in 1960, while maintaining his own New York office, Michael Lax Associates Inc., founded in 1965. He was awarded the Prix de Rome in Design in 1978. In 1984, he established a sculpture workshop in Italy, and in 1994, one in New York.

Robert Lepper (1906-1991) Artist, sculptor and design educator who was born in Aspinwall, Pennsylvania, graduated from Carnegie Institute of Technology (CIT) in 1927 and spent a year in Europe absorbing art and Bauhaus design. He joined the faculty at CIT in 1930, where he painted machine art and was a key participant in the first degree-granting program of industrial design initiated in 1934. In 1938 he developed and published *The Elements of Visual Perception,* linking art elements to manufacturing processes. He taught in the industrial design program from its inception until his retirement in 1975.

Monte Levin (b. 1917) U.S. industrial designer who was in the first fully-matriculated Pratt Institute design class of 1938, served in the U.S. Army as a combat engineer captain during World War II, and opened his New York office, Monte Levin Associates, Inc. in 1945. He designed a total of over 8000 products for 150 clients, and after 1968, designed most of Hamilton Beach's products.

J. Gordon Lippincott (1909-1998). U.S. industrial designer who graduated from Swarthmore College in 1931 as an engineer, and obtained a master's degree in architecture and civil engineering at Columbia University. He joined the faculty at Pratt in 1936 with Donald Dohner, assisted in the establishment of its design education program, and began consulting as an industrial designer while teaching. He wrote *Economics of Design* in 1937. He opened a design office in New York with Donald Dohner in 1943, Dohner & Lippincott, in partnership with the Douglas T. Sterling Company of Stamford, Connecticut. Upon Dohner's death, he continued editing of the industrial design section of *Interiors* magazine. The firm became J. Gordon Lippincott and Company in 1944. In 1944 Walter Margulies joined him and the firm was renamed Lippincott & Margulies in 1947. He wrote *Design for Business,* published by Paul Theobald in 1947. Lippincott retired in 1969, selling his interests to Margulies, and engaged in travel, investments, and consulting. The firm continues today as Lippincott Mercer.

Raymond Loewy (1893-1986). Probably the best known industrial designer, he was born in Paris, studied engineering, served in the French army in World War I, and arrived in the U.S. in 1919. He worked as a fashion illustrator and costume designer before opening his office in New York in 1929. He was one of the founders of SID in 1944. He was on the cover of *Time* magazine in 1949, wrote *Never Leave Well Enough Alone* in 1951, and established a Paris office in 1952. By 1960 his office, Raymond Loewy Associates (RLA), had a staff of 180. Loewy's major designs included the 1932 and 1934 Hupmobiles, the 1934 GG-1 locomotive for Pennsylvania Railroad, the 1947 and 1953 Studebakers, the 1962 *Avanti*, the 1966 Exxon logo, the 1970 U.S. Post Office logo, *Air Force One* livery and interiors in 1973, NASA Skylab 2 space station interiors in 1973, and *Concorde* interiors in 1976. His New York office closed in 1977. Primary Loewy archives are at the U.S. Library of Congress and the Hagley Museum & Library in Wilmington, Delaware.

Adolph Loos (1870-1933). A writer who was born in Czechoslovakia, he studied building at Liberec (1887-88) and then at Dresden Institute of Technology in Germany. He worked in the U.S. from 1893 to 1896 as an architect. Returning to Vienna, he set up an architectural practice and designed furniture. His greatest influence was in his writings, particularly *Ornament and Crime*, but he founded a review, *Das Andere* (The Other) in 1903, in which he wrote of his admiration for British and American simplicity. His Steiner House (1910) set a precedent and standard for the removal of all decoration, and greatly influenced Le Corbusier, Erich Mendelsohn, and Walter Gropius.

Alvin Lustig (1915-1955) U.S. graphic and interior designer who was born in Colorado, trained at the Art Center School in Los Angeles, and studied briefly with Frank Lloyd Wright. He opened his office in 1936 in Los Angeles. During 1935-36 he was design director for *Look* magazine and he headed advertising for Knoll furniture prior to 1946. In 1951 he opened an office in New York, concentrating on book and magazine design. But his eyesight began to fail and he was totally blind by late 1954. Still, he continued to teach and design until his death.

Charles Rennie Mackintosh (1868-1928). Scottish architect, designer, and foremost exponent of Art Nouveau in Britain, he trained at Glasgow School of Art and apprenticed in an architectural office in Glasgow in 1889. He traveled to Europe in the 1890s, and was influenced by Celtic designs. His posters were exhibited in the 1896 Arts and Crafts Exhibition. His wife Margaret Macdonald, her sister Frances, and her husband J. Herbert MacNair were his closest associates in Glasgow and known as "The Four." His best-known work is the Glasgow School of Art building, for which he won a competition in 1897. It was completed in 1909. He moved to London in 1914, and after 1923 devoted himself to painting.

Tucker Madawick (b. 1917). U.S. industrial and automotive designer who was born in New York City, attended Brooklyn Technical High School, the Art Students League, and Pratt Institute. He joined Ford Motor Company in 1939, where he worked on Mercurys and Lincoln *Continentals*. In 1942 he joined Ford's aircraft team and from 1943 to 1946 worked at the B-24 bomber facility at Willow Run, Michigan, and later as production coordinator for the Convair B-32 and B-36 bomber programs. In 1946 he joined J. Gordon Lippincott and was involved with the Tucker car. He joined Raymond Loewy Associates in 1947 and re-established Loewy's London office. In 1950 he joined Loewy's team in South Bend, Indiana, to work on the 1953 Studebaker *Starliner*. He joined RCA in 1959 as manager of radio, phonograph, tape recorder, and television design, becoming vice-president of the Advanced Design Center in 1968 and divisional vice-president of Consumer Electronics from 1971 until his retirement in 1980. He was president of IDI in 1964.

Robert Maillart (1872-1940) Swiss engineer who was among first to work with reinforced concrete, a material he treated with economy, grace and remarkably good form. Born in Bern, Switzerland he studied engineering at Swiss Federal Polytechnical Institute in Zurich. He developed his unique "hollow box" construction principle for the Zuoz bridge when working for the Zurich construction firm, Froté and Westermann. He set up own construction firm in 1902, and completed his first masterpiece, the Tavasana Bridge, in 1905. He continued to refine his designs with the Valtschielbach Bridge (1925), the Salginatobel Bridge (1930), the Schwanbach Bridge (1933), and the Vessy bridge (1936), all in Switzerland.

Kasimir Malevich (1878-1935). Russian artist who, along with Piet Mondrian and Wassily Kandinsky, was one of the three founding fathers of 20th century abstract painting. He studied at Kiev Drawing School (1895-1896) and the Moscow School of Art. He was influenced by *Cubism* and *Futurism* (1910-1920), and in 1915, launched *Suprematism*. After the Revolution, he began teaching at the First State Free Art Studios in Moscow, a forerunner of VKHUTEMAS, then replaced Marc Chagall as director of the Vitebsk Institute of Art and Practical Work in 1919. In 1920, he founded the UNOVIS group that included El Lissitsky and Nikolai Suetin. After 1919 he turned from painting to the design of cities and housing. In 1927 he visited the Bauhaus in Dessau, which revived his ideas. He disseminated them in a book, *The Non-Objective World.*

Ellen B. Manderfield (1916-1999) U.S. industrial designer who received a BA in Art from Mundelein College of Loyola University in Chicago and began her career with the Colonial Radio Company (later known as Sylvania) in Buffalo, New York from 1944 to 1947. She worked in the Montgomery Ward design group from 1947 to 1951, and for General Electric Company in Syracuse, New York from 1951 to 1956. She joined Oneida Ltd. in 1956, and before her retirement in 1986, over 200 of her flatware designs were produced, many of which are still on the market. She was the first female member of ASID.

Walter Margulies (1914-1986) U.S. architect and interior designer who worked for the Schine Company chain of hotels and theatres. He joined with J. Gordon Lippincott in 1944 who re-named his firm Lippincott & Margulies in 1947. Soon after 1947, he married the daughter of Henry A. Wallace, Franklin Roosevelt's vice-president (1941-1945). Upon Lippincott's retirement in 1969, he purchased Lippincott's interests and headed the firm until 1982, when he sold it to Clive Chajet. The firm is now known as Lippincott Mercer.

Fritz Mayhew U.S. automotive designer who was born in Pittsburgh, Pennsylvania, graduated with a BFA in industrial design from Carnegie Mellon University in 1963, and was selected as IDSA's Outstanding College Graduate of that year. He joined Ford Motor Company design staff and led the design team for the 1986 Ford *Taurus*. As executive director for North American Design through the 1990s, he was responsible for all Ford cars and trucks including the retro 2002 *Thunderbird*. A fine artist, as well, his paintings have been exhibited widely in the South and Midwest, where he has had 15 one-man shows. Since 1998, he divides his time between his studios in Harbor Springs, Michigan and Naples, Florida.

J Mays (b. 1954) U.S. automotive designer who was born in Pauls Valley, Oklahoma, went to University of Oklahoma, and then to the Art Center College of Design in Pasadena, graduating in 1980. He joined Audi and designed the 1991 *Audi Avus* concept car. In 1989 he became head of Volkswagen's U.S. design center, where he and Freeman Thomas designed the *Concept One* retro-*Beetle*, unveiled in 1994 and introduced in 1998 as the new *Beetle*. In 1995 he became vice president for design development at SHR Perceptual Management in Scottsdale, Arizona and in 1997 became corporate vice president of design at Ford.

Paul McCobb (1917-1969) U.S. furniture designer who studied painting at Vesper George School of Art in Boston and began work as a muralist, then as interior designer with Jordan Marsh. He established his own New York firm in 1945. He formed an organization in 1950 with B. G. Mesberg as distributor and Winchendon Furniture Company as manufacturer to produce his *Planner Group* furniture that was distributed in department stores across the country. *Business Week* magazine called him the "creator of the contemporary look for young moderns."

Katherine J. McCoy (b. 1945) U.S. designer and educator who was born in Decatur, Illinois, studied industrial design at Michigan State University before joining Unimark International in 1967 and went on to work at Chrysler Corporation and Omnigraphics Inc. In 1971, She became co-chair, with her husband Mike McCoy, of the design department at Cranbrook Academy of Art, which they continued to direct until 1995, and concurrently maintained their design consultancy, McCoy & McCoy. She is past president and Fellow of the Industrial Designers Society of America, and was president of the American Center for Design and vice president of the American Institute of Graphic Arts.

Donald L. McFarland U.S. industrial designer who received a BS in aeronautical engineering from Rensselaer Polytechnic Institute and worked at Chance Vought aircraft to create the first carrier jet fighter. In 1951 he became manager of product planning for General Electric in Bridgeport, Connecticut and in 1957, head of all small appliances. He was president of ASID in 1958. In 1960 he joined with Latham Tyler Jensen, Inc. as a partner and established its West Coast office, and in 1973 established McFarland Design, Inc. in Santa Barbara, California.

Duncan McRae (1919-1984) U.S. automotive designer who was born in Lansing, Michigan and started his career with Ford in 1938. In 1942 he joined Chrysler, served in the Army Air Corps 1943-1945, and returned to Chrysler after the war. From 1947 to 1949 he worked with Kaiser-Frazer, from 1949 to 1952 with Ford, from 1955 to 1960 with Studebaker- Packard, from 1964 to 1974 with Ford, and then then retired to Australia.

Erich Mendelsohn (1887-1953) German architect who trained at the Technical High School in Munich, was active in theatre design, and well known in the U.S. from 1915 onward due to extensive publications and exhibits showing his buildings with "streamlining" characteristics of fluid, rounded shapes. He opened his own office in Berlin in 1919, and had a major New York exhibition in 1929. In 1933 he left Germany for England, where he established a partnership with Serge Chermayeff. He worked in Palestine from 1939 until 1941, when he emigrated to the U.S., became a citizen in 1946, and worked in San Francisco and St. Paul, Minnesota.

Adolph Meyer (1881-1929). German architect who was born in Mechernich-Eifel. studied at Düsseldorf Arts and Crafts School and was employed 1907-1908 by Peter Behrens and by Bruno Paul 1909-1910. He was Walter Gropius's closest collaborator until 1925 and thereafter directed a consulting office for city planning in Frankfurt, Germany and taught an architecture class at the art school there.

Hannes Meyer (1889-1954) Swiss architect and anthroposophist who worked 1909-1912 for land reform in Berlin, and traveled in Britain 1912-1913 to study English garden cities. In 1927 he was appointed head of the new architecture department at the Bauhaus, assisting Walter Gropius with an experimental housing project at Törten. Because his Marxist leanings so encouraged opponents of the school, he was forced to resign in 1930. He went to Russia until 1936, then returned to Switzerland. From 1939 to 1949 he was director of an institute for town planning in Mexico.

Ludwig Mies van der Rohe (1886-1969). A German architect and furniture designer, who is famed for his quote "Less is more," and his steel and glass International Style architecture, now known as "Miesian". He was born in Germany, trained as a stonemason in his father's business, and apprenticed with furniture designer Bruno Paul. He trained with Peter Behrens from 1908 until 1912, when he established the office he maintained until 1938. He was a member of the *Deutsche Werkbund* starting in 1921. He designed the German pavilion and his famous chairs for the Barcelona International Exhibition in 1927, as well as a cantilever steel chair of that era. He was the third, and last, director of the Bauhaus from 1930 to 1933. He emigrated to the U.S. in 1938, designed buildings for the Illinois Institute of Technology (Armour Institute) until 1940, and became its director of architecture in 1948. He designed the Seagram Building in New York in 1958.

William Mitchell (1912-1988) U.S. automotive designer who was born in Cleveland, studied art at Carnegie Institute of Technology and at the Art Students League, and worked briefly as a commercial artist. He joined General Motors in 1935 and worked in the Cadillac studio where he designed the Cadillac script logotype and the 1938 Cadillac *60 Special*. As the protégé of Harley Earl, he became vice president of styling in 1958, upon the announcement of Earl's retirement in 1959. Mitchell retired in 1977.

Bill Moggridge (b. 1943) U.S industrial designer born in the UK, graduated from the Royal College of Art, and worked as a designer with The Hoover Company, Ltd. until he established his own office, Moggridge Associates, in London in 1969. In 1979, he emigrated to the U.S. and established ID Two in Palo Alto, California, with compatriot, Mike Nuttal. In 1980, he designed the first laptop computer, the GriD *Compass*. In 1990, ID Two merged with Matrix (started by Nuttall in 1983) and David Kelley to form IDEO. By 1993, IDEO was the largest and most successful design office in the U.S.

László Moholy-Nagy (1895-1946). Hungarian painter, photographer, filmmaker, typographer, and educator who studied law in Budapest and later served in World War I. After the war, he began painting and experimented with photography in Berlin. He joined the Bauhaus in 1923, where he ran the metal workshop until his departure in 1928. In 1929, he wrote *Von Material zu Arckitektur,* which was translated in 1932 as *The New Vision.* He worked in Berlin, Amsterdam, and, from 1935 to 1937, in London. In 1937 he came to Chicago at the invitation of Walter Gropius to become director of the school of design at the Association of Arts and Industries in Chicago, which he promptly renamed "The New Bauhaus: American School of Design." He concurrently founded the Chicago Institute of Design Research, with Sigfried Gideon as director, and wrote *Vision in Motion* in 1946.

Piet Mondrian (1872-1944) Artist who was born Pieter Cornelis Mondriaan near Amsterdam, the Netherlands and along with Kasimir Malevich and Wassily Kandinsky, was one of the three founding fathers of 20th century abstract painting. He took drawing lessons from his father, a lithographer, and studied two years at Rijksacademie in Amsterdam. He went to Paris in 1912 where he took up cubism, but returned in 1914 to Amsterdam to explore abstraction. In 1919 he returned to Paris to write essays for *De Stijl*. In 1938 he went to England as war approached, and on to New York in 1940.

John Richard "Jack" Morgan (1903-1986) U.S. industrial designer who was born in Guatemala City and was with Harley Earl's Art & Color group at General Motors in 1929. He headed Sears, Roebucks' first industrial design department and began a private practice starting in 1935. His *Waterwitch* outboard motor for Sears was in the 1934 *Machine Art* Show. With Sears until 1943, he designed a number of products including a wash machine, upright vacuum cleaner, and kitchen cabinet ensemble. From 1947 to 1952 he was a partner with Mel Boldt in Chicago, and from 1948 was a design consultant for Dormeyer Corporation housewares.

Koloman Moser (1868-1918) Austrian artist, graphic and furniture designer who studied in Vienna at the Academy (1886-1892) and at the School of Arts and Crafts (1892-1895), where he taught from 1899 onward. His graphic design progressed from organic to geometric, typical of the *Wiener Sezession* (Vienna Secession), of which he was a co-founder. He was also a co-founder of the *Wiener Werkstätte* for which he created many designs

Peter Müller-Munk (1904-1967). Industrial designer who was born in Germany, studied as a silversmith at the University of Berlin, and emigrated to the U.S. in 1926. He worked at Tiffany's in New York as a metalworker (1926-1929), then established his own silver studio in 1929, exhibiting in New York shows. In 1935 he accepted a position at Carnegie Institute of Technology in Pittsburgh as head of the industrial design program. He remained there until 1944, when he left to devote his time fully to the industrial design office he had established in Pittsburgh in 1938. Major clients included Waring for which he designed its classic blender and, in 1943, the Dow Chemical Company, for which he designed concepts to stimulate postwar public interest in plastics. In 1954 he was president of the Society of Industrial Designers, and, in 1957, the first president of the International Council of Societies of Industrial Design (ICSID).

Hermann Muthesius (1861-1927). German architect, writer, and spokesman for design, who in 1893 became the official architect of the Prussian government, where he helped reform schools of design. In 1896 he was appointed to the German Embassy in London to study and report on English design developments. There he met Charles Rennie Mackintosh and published numerous articles in *Dekorativ Kunst* on the Arts and Crafts movement. In 1903, he was appointed to the Prussian Ministry of Trade and Commerce in charge of art education. He published *Das Englische Haus* in 1904, and urged the establishment of the *Deutsche Werkbund in* 1907.

C. Stowe Myers (1906-1995) U.S. industrial designer who was born in Altoona, Pennsylvania, and received a BA in architecture from the University of Pennsylvania in 1930. Norman Bel Geddes hired him in 1932 to create illustrations for his book, *Horizons.* He joined Walter Dorwin Teague in 1934, was named a partner in 1945, and headed Teague's Los Angeles office until it closed in 1949. From 1950 to 1952, he managed Loewy's Chicago office and from 1954 to 1970, headed Stowe Myers Design in Evanston, Illinois. He retired in 1976 to paint, and was named official artist of the U.S. Air Force in 1978. He moved to Louisiana in 1988.

George Nelson (1908-1986). U.S. architect, furniture designer, author, and one of the most influential figures in post-war U.S. design. He was born in Hartford, Connecticut, graduated from Yale in 1931, and, while doing post-graduate work at Catholic University, Washington, DC, was awarded the Rome Prize in architecture in 1932. He became assistant editor of *Architectural Forum* in 1933, and co-managing editor with Henry Wright in 1943. He established an office in 1936 in association with William Hamby, which continued until after the war. In 1945, he and Wright developed the unique concept of storage wall systems. From 1946 to 1965, Nelson directed design for the Herman Miller Furniture Company. In 1947 he opened his own design studio while continuing as a consultant with Herman Miller. He wrote *Tomorrow's House* (1945) with Henry Wright, *Problems of Design* (1957), and *How to See* (1977).

Marcello Nizzoli (1887-1969). Italian industrial designer who studied painting, architecture, and graphics at the Academy of Fine Arts in Parma (1910-1913). He established a studio in Milan, working with Giuseppe Terragni (1931-1936). In 1936 he became chief consultant designer to Olivetti, and was hired full-time in 1938. He designed the *Lettera 22* typewriter (1950), the *Lexicon 80* (1948), and headed design for Olivetti until the 1960s.

Isamu Noguchi (1904-1988). U.S. sculptor and designer who was born in Los Angeles and spent his childhood in Japan. He returned to the U.S. in 1918 to study medicine, but abandoned it to study at the Leonardo da Vinci Art School in New York, under Onorio Ruotolo (1922-1926). He studied sculpture in Paris (1927-1928) with Constantin Brancusi before making pottery in Kyoto. He returned to the U.S. in 1931, where he built his reputation as a well-known public works sculptor, settling in 1960 in Long Island City, New York, which became the site of the Isamu Noguchi Garden Museum after his death. His important product designs include a glass-topped sculptural coffee table for the Herman Miller Furniture Company and the *Akari* lamp for Knoll.

Eliot Fette Noyes (1910-1977). U.S. architect and industrial designer who was born in Boston, Massachusetts, studied architecture at Harvard (1932-1938), and was employed by Marcel Breuer and Walter Gropius after graduation. He worked as a curator at the Museum of Modern Art, and was appointed director of its new department of industrial design in 1940. During World War II he served as a major in the Air Force. After the war he served as design director for Norman Bel Geddes until 1947, when he opened his own office. Thomas J. Watson, Jr., of IBM, retained him in 1948 for product design, and in 1956 to develop a unique corporate style similar to Olivetti. He did so with the help of Paul Rand, Marcel Breuer, and Charles Eames. In 1960, he was retained by Westinghouse to repair the company's deteriorating image, again using Paul Rand for graphic design. He died in New Canaan, Connecticut.

Joe Oros U.S. automotive designer who graduated from Cleveland Institute of Art in 1939 and started his career with George Walker, where he worked on the 1940 Lincoln *Continental* and the 1949 Ford. His wife and college classmate, Elizabeth Thatcher Oros, was the first woman to work in an automotive studio, on the 1941 Hudson. Joe joined Ford full-time in 1955 when Walker was named head, and became head of Ford's styling studio from 1956 until 1968, when he became vice president of Ford Europe. In 1974, he became executive director for Lincoln-Mercury until his retirement in 1975.

Carl Otto (d. 1986). U.S industrial designer who was born in Michigan and trained as an engineer at Michigan State. He began his career as a draftsman with Duesenberg, then came to New York to work for Norman Bel Geddes. In 1935 he joined Raymond Loewy to work on Pennsylvania Railroad designs. He opened Loewy's pre-war office in London in 1938, which he headed until 1940, when he became an associate. One of his accounts was Standard Auto (Triumph). He resigned from Loewy's office in 1951, returned to the U.S., and opened his own firm with offices in New York, Stockholm, and London, which operated until 1954. His best-known designs include the *Schick 20* razor and the Edison *Voicewriter.*

J.J.P. Oud (1890-1963) Dutch modernist architect born Jacobus Johannes Pieter Oud who was trained in Amsterdam and Delft, worked briefly in Germany, and returned to practice in Leyden, The Netherlands. He met Theo van Doesburg and Gerrit Rietveld in 1915, and became associated with the *De Stijl* movement. He became Rotterdam's city architect in 1918. Best known examples of his work are the Houses at the Hook of Holland (1926-1927) and the Kiefhoeck project in Rotterdam.

Verner Panton (1926-1998) Danish industrial designer who studied at the Royal Danish Academy of Fine Arts, Copenhagen from 1947 to1951, and became an associate of Arne Jacobsen from 1950 to 1952. He went on to establish his own architectural and design office in Denmark from 1955 to 1962 and in 1963 moved his work to Binningen, Switzerland. He is known for his one-piece plastic chair of 1967 and *Cone* chair of 1959.

Victor J. Papanek (1925-1998) U.S. industrial designer, anthropologist, writer, and teacher who was born in Vienna, Austria, arrived in the U.S. in 1932, and graduated in architecture and industrial design from the Cooper Union in 1948. He opened his consulting office in 1953 and was dean of the School of Design at California Institute of the Arts. In the early 1970s he became chairman of design at the Kansas City, Missouri, Art Institute and wrote a number of books, including *Design for the Real World* (1972), and *How Things Don't Work*, (1977). From 1981, he taught architecture and design at the University of Kansas.

Joseph Marshall Parriott (1920-2000) U.S. industrial designer and design educator who was born in Moundsville, West Virginia and raised in Colorado. He attended Colorado University and Pratt Institute, and served with the U.S. Army in World War II. He worked for Raymond Loewy/ William Snaith as manager of product design, and then became a consultant for Becker & Becker. In 1965, as an officer of IDI, he participated in the formation of IDSA in 1965 and was its president in 1966. He also became chairman of Pratt's industrial design department in 1966, a post he held for 24 years. He retired in 1992 to Cooperstown, New York.

Raymond E. Patten (1897-1948) U.S. industrial designer born in Malden, Massachusetts who attended Harvard University and studied engineering at the Massachusetts Institute of Technology. He started as an automotive designer for several companies including Packard, and in 1928 became a consulting designer for the Edison General Electric Company (GE) in Chicago to supervise appearance design. In 1933-1934 he designed exhibits for GE at the *Century of Progress Exhibition*. In 1933, GE established an internal industrial design group in Bridgeport, Connecticut, and Patten was named as head. He relocated there in 1934 and held the position until his death. He designed ranges, dishwashers, and locomotives, and was a charter member of the Society of Industrial Designers when it was formed in 1944.

Giovanni (Gio) Ponti (1891-1979) Italian architect, designer and champion of industrial design who was born in Milan, and studied architecture at the Polytechnic in Milan. He helped organize the 1923 *Biennale* before establishing an architectural firm with Emilio Lancia in 1927. In 1928, he founded and edited *Domus,* the magazine serving as the mouthpiece for the *Novecento* group he founded in 1926, which had brought the influence of the *Wiener Werkstätte* to Italy. He was a professor at Polytechnic from 1936 to 1961, edited *Stile*, a design magazine, from 1940 to 1947, and helped organize the *Compasso d'Oro* in 1954.

Arthur Jon Pulos (1917-1997). U.S. industrial designer, educator, and design historian who was born in North Vandergrift, Pennsylvania, graduated in art education from Carnegie Institute of Technology in 1939, and earned a Masters in silversmithing at the University of Oregon in 1943. He served in the U.S. Air Force during World War II, was chairman of the silversmith department at the University of Oregon, and became associate professor of design at the University of Illinois from 1946 to 1955. He went to Syracuse University in 1955 to direct the industrial design program, becoming chairman of the Department of Design in 1970 until his retirement in 1982. He founded Pulos Design Associates in 1958 and remained in active practice until 1988. He was the last president of IDEA in 1964, president of IDSA 1973-1974, and president of ICSID 1979-1981. He wrote *American Design Ethic* (1983) and *American Design Adventure* (1988).

Jens Quistgaard (b. 1919). Danish designer and son of a sculptor father who soon learned to use the tools of craftsmanship and before World War II served as an apprentice at the Georg Jensen Smithy in Copenhagen. After the war he became a free-lance designer and won a number of design awards. In 1954, with U.S. entrepreneur/engineer Ted Nierenberg, he established Dansk, and designed many of its most successful products over the next 20 years.

Dieter Rams (b. 1932). German industrial designer who studied architecture and design at Wiesbaden Werk-Kunstschule (1947-1953). He started with Braun AG in 1955, and in 1957 was appointed head of its design program, which earned worldwide acclaim. After 1957 he designed furniture for Otto Zapf, for Vitsoe + Zapf, and for Niels Weise Vitsoe. He remained in a lead role at Braun as chairman of the board.

Paul Rand (1914-1996). U.S. graphic designer who studied at Pratt Institute (1929-1932), at Parsons (1932-1933), and at Art Students League under George Grosz (1893-1959). By 1938 he was art director of *Esquire-Coronet* and *Apparel Arts* magazines. He opened his office in 1939. From 1941 to 1954, he was art director at the William H. Weintraub Agency, New York. He wrote *Thoughts on Design* in 1947. He consulted with IBM starting in 1956, designing IBM's trademark and what was probably its first corporate design manual. He also designed logos for ABC and UPS. He died in Norwalk, Connecticut.

William E. Reddig U.S. industrial designer who was a 1949 graduate of Pratt Institute, joined Nash-Kelvinator in 1950, and from 1958 to 1968 headed Kelvinator design. From 1968 to 1973 he was vice president of engineering at Kelvinator, and from 1973 until his retirement in 1983, was president of GR Manufacturing Company, a division of White Consolidated Industries. He maintained Reddig Associates from 1983 to 1986, and from 1986 to 1991, was president of Curtis Wire Products in Petoskey, Michigan.

Lilly Reich (1885-1947) German industrial designer who was the first woman to become a member of the *Deutsches Werkbund's* board of directors in 1920. She collaborated with Mies van der Rohe in numerous architectural and product design projects and directed the weaving studio and interiors workshops at the Bauhaus in 1933.

Jean Reinecke (1909-1987). U.S. industrial designer who was born in Bourbon County, Kansas, attended Kansas State Teachers College, and studied art at Washington University in St. Louis, Missouri. He started as a display artist at General Display Studios in St. Louis, soon becoming a partner. In 1930 he opened a Chicago office to work on displays for the Chicago *Century of Progress Exposition* (1933-1934). In 1934 he joined with James Barnes to form Barnes & Reinecke. In 1938 he designed the *Toastmaster* model 1B9 for McGraw Electric Company (later McGraw-Edison), the first to exploit the curvature of chrome shells. He designed Scotch tape dispensers for 3M starting in 1938, including its ubiquitous 1940 plastic disposable dispenser. In 1948, he sold his interest in the partnership and established J.O Reinecke & Associates, with offices in Chicago and, later, Pasadena, California, where he designed until 1986. He was president of SID in 1952.

Frederick Hurten Rhead (1883-1942) U.S. ceramics designer who was born in England, emigrated to the U.S. in 1902 to work in Ohio and in 1904 became art director of the Roseville pottery there. In 1908 he joined the William Jervis Pottery on Long Island, and in 1909 was a pottery instructor in St. Louis. From 1911 to 1913 he was with the Arequipa Pottery in California and in 1916 published an industry magazine, *The Potter*. He returned to Ohio in 1917 and in 1927 joined the Homer Laughlin Company, where he designed *Fiesta* in 1936, *Harlequin* in 1938, and headed the design department until his death.

Deane W. Richardson (b. 1930). U.S. industrial designer who was an industrial design graduate of Pratt Institute. With David B. Smith (b. 1932), he co-founded Richardson/Smith (R/S) in Columbus, Ohio, in 1959 to capitalize on a central Midwest location. R/S correctly foresaw dramatic changes within the profession, corporate structures, and global design markets, and pioneered a participative management style and multidisciplinary staff environment for design offices that later became common practice. By 1985 R/S had a staff of over 150. In 1988 R/S became a subsidiary of Fitch & Company PLC, founded in 1971 by Rodney Fitch, a British architect. The new firm was known as Fitch Richardson Smith, Inc., with a combined staff of 500, and in 1993 became known as Fitch, Inc. Deane Richardson was president of ICSID 1992-1993.

John Gordon Rideout (1898-1951). U.S. industrial designer, born in St. Paul, Minnesota. He studied at St. Paul Academy and in architecture at the University of Washington, but left school to serve in the Navy in World War I. He worked in the 1920s as a graphic designer in Chicago, where he was one of the founders of the Chicago Society of Typographic Arts. In 1931 he went to Toledo, Ohio, to partner with Harold Van Doren in a design office. Rideout left the firm in 1935 to start his own office in Cleveland, Ohio, John Gordon Rideout and Staff. World War II in 1941 closed his office, but he re-opened it in 1943 in Chagrin Falls, Ohio. In 1944 he joined with architect Ernest Payer to combine industrial design and architecture. Their relationship ended in 1946. Rideout was a co-founder of SID.

Richard Riemerschmid (1868-1957) German designer associated with the *Wiener Werkstätte* who studied painting at the Academy in Munich. He worked in Art Nouveau style but sought functional solutions. He taught in Nuremburg from 1902 to 1905, worked in the *Dresden Workshops* from 1903 and was a founding member of the *Deutche Werkbund* in 1907. He was director of the Munich School of Applied Arts from 1912 to 1924, and headed the Werkschule in Cologne from 1926 to 1931.

Gerrit Thomas Rietveld (1888-1964) Dutch architect and designer who trained as a cabinetmaker apprentice in his father's workshop and worked for goldsmith Casrel Begeer from 1906 to 1911. He designed the *Red/Blue Chair* in 1918 and became a member of the *De Stijl* movement in 1919. In 1924 he designed and built a house with Truus Schröder. He participated in the founding of the Congrés Internationaux D'Architecture Moderne in 1928 and continued to practice architecture until his death.

Jens Risom (b.1916) U.S. furniture designer who was born in Denmark and arrived in the U.S. in 1938. He worked as a textile and furniture designer with Dan Cooper in 1939, joined with Hans Knoll in 1941 to develop a chair which became the basis for Knoll's first line of products in 1942. He joined the U.S. Army in 1943 and started his own design and manufacturing business, Jens Risom Design, in 1946.

Aleksandr Rodchenko (1891-1956) A leading proponent of Russian *Constructivism* who studied at the Art School in Kazan from 1910 to 1914. He moved to Moscow and became acquainted with Kasimir Malevich and Vladimir Tatlin. With his wife, Varvarya Stepanova, and Alekseye Gan, he formed the first working group of the *Constructivists* in 1921. From 1923 on he designed posters for government trading organizations, the cinema, the journal LEF (until 1925), and then Novyi LEF from 1927 to 1929.

Gilbert Rohde (1894-1944). U.S. furniture designer who was born in New York, and studied at the Art Students League and Grand Central School of Art in New York. He became a drama and music critic, cartoonist, reporter, and furniture illustrator. He opened his office in New York in 1927 designing furniture and showrooms. In 1932 he was appointed director of design for the Herman Miller Furniture Company in Zeeland, Michigan, where he served until his death. He was a member of the American Union of Decorative Artists and Craftsmen (AUDAC) in the early 1930s, one of the founders of the Designers Institute of the American Furniture Mart in 1933, director of the Design Laboratory School from 1935 to 1938, and headed an industrial design program at New York University's School of Architecture from 1939 to 1943. He designed major exhibits for the New York World's Fair in 1939.

Irvin W. Rybicki. (b. 1924) U.S. automotive designer who joined General Motors' design staff in 1945, and became group chief designer for Cadillac, Buick, Oldsmobile by the end of the 1960s. He replaced retiring William L. Mitchell in 1977.

Eero Saarinen (1910-1961). Architect and furniture designer who was born in Kirkkonummi, Finland, the son of Eliel Saarinen, Finland's leading architect and first president of Cranbrook Academy of Art when it was founded in 1925. He emigrated to the U.S. in 1923 with his parents, studied sculpture at Academie de la Grande Chaumiere in Paris (1929-1930), and received a degree in architecture from Yale University (1930-1934). He traveled and studied in Europe, the Middle East, and Africa on a fellowship (1934-1936). In 1935 he joined his father at Cranbrook, where he worked with Charles Eames, who was hired as head of experimental design there in 1937. They explored the use of new plastics in furniture design together. Eero taught architecture at Cranbrook from 1939 to 1941, and continued his father's practice as Eero Saarinen Associates after 1950. He married Aline B. Louchheim, who became well known as articulate art critic Aline Saarinen. With Charles Eames, Eero won first prize in a 1941 furniture design competition sponsored by the Museum of Modern Art. He joined Knoll Associates in 1947, and designed its 1948 *Womb* chair and its 1956 line of *Pedestal* furniture. He is also well known for his architectural design of the TWA Terminal at John F. Kennedy airport (1962), Dulles Air Terminal (1958), and the Gateway Arch in St. Louis (1965).

Eliel Saarinen (1873-1950). Finnish architect who began his practice in 1896 in Helsinki in partnership with Herman Gesellius and Armas Lindgren and is known as Finland's "National Architect." In 1923 he won second prize for design of Chicago Tribune Tower design competition (Raymond M. Hood and John Mead Howells won the first prize) and emigrated to the U.S. He joined the staff at the University of Michigan, and then co-founded Cranbrook Academy of Art, becoming its president in 1932. He brought Charles Eames, Harry Bertoia, and his son, Eero Saarinen together on the teaching staff.

Alexis Wladimirovich de Sakhnoffsky (1901-1964) Automotive designer who was born in Russia, studied engineering in 1914 at the University of Lausanne, Switzerland, and arts and crafts at the Ecole des Arts et Métiers in Brussels starting in 1920. He began work for coachbuilder Van den Plas in Belgium, and became its art director in 1924. In 1928, as art director for the Hayes Body Company in Grand Rapids, Michigan, he designed the 1929 American Austin. When Hayes failed in 1931, he was recruited by Auburn and became known as an outspoken futurist and stylist. In 1963, he executed a portfolio of classic car drawings for *Automobile Quarterly*.

George Sakier (1897-1988) U.S. industrial designer who was born in France, became interested in art and emigrated to the U.S. in the early 1920s. As an engineer in New York, he designed machinery and taught machine design while writing and practicing in fine art. He exhibited his landscape art in galleries, and designed glassware for Fostoria Glass, Moundsville, West Virginia for over 50 years. By 1927, while continuing his independent practice, he headed the Bureau of Design Development at the American Radiator & Standard Sanitary Corporation. He was featured in the important 1934 *Fortune* article, and designed exhibits for the 1939 World's Fair.

Clair Samhammer (b. c. 1929) U.S. industrial designer who studied at Pratt Institute from 1947 to 1951. He worked in the early 1950s for Sundberg-Ferar, then with Harley Earl Inc. in Detroit, Michigan, where he designed a toy *Corvette* pedal car. In 1959, he joined Melvin Best Associates in Pasadena, California and in 1962, joined Design West. His best-known designs are the award-winning attaché cases for Sampsonite Corporation, the *Classic* (1962) and the *Royal Traveler* (1965). In 1981, he partnered with Bob Green to establish the Creative Design Center in Orange, California.

Raymond C. Sandin (d. c. 1986) U.S. industrial designer who was born in Sweden, came to the U.S. at age 20 and studied architecture at Illinois Institute of Technology before joining Hotpoint Division of General Electric in 1935 as a one-man design department. During World War II he offered Hotpoint service to help customers plan kitchens, which gathered information valuable in post-war products. He headed Hotpoint design for the remainder of his career and in 1965, his staff numbered 13.

Antonio Sant'Elia (1888-1916) Italian architect who was born in Como, studied in Milan and at the University of Bologna and established himself as an architect in Milan in 1912. During 1912-1914 he made a number of drawings of building and town planning shown in 1914 in an exhibition, *Cittá Nuova* (New City) and his work became the most publicized expression of the *Futurists.* He entered the Army before Italy entered World War I, and was killed in the battle of Monfalcone

Richard Sapper (b. 1932) German industrial designer who studied engineering in Munich and worked in Germany for Daimler-Benz from 1956 to 1958, when he moved to Milan, worked for Gio Ponti, then for La Rinascente department store. He collaborated with Marco Zanuso from 1960 to 1977, a partnership responsible for a number of high-tech electronic products for BrionVega and the 1967 award-winning *Grillo* telephone. In 1972, Sapper designed the *Tizio* lamp for Artemide and the 1983 *Bollitore* kettle for Alessi. From 1980 he was a design consultant for IBM, and from 1986 he has been a professor at the Stuttgart Art Academy.

Samuel Scherr (1923-2002) U.S. industrial designer who was born in Akron, Ohio, and was a 1946 industrial design graduate of the Cleveland Institute of Art. He spent a year with General Motors, then joined with his classmate, Eugene Smith and Bernard McDermott to form the Akron firm of Smith, Scherr & McDermott in 1947. In 1960, the firm became Scherr & McDermott, with Scherr at President and CEO. After its close in 1975 he became president of The American Crafts Council and American Crafts Museum in New York, and later relocated to North Carolina with his wife, noted jewelry designer Mary Ann Scherr.

Peter Schlumbohm (1896-1962) German inventor and chemist who earned a doctorate in chemistry from the University of Berlin before coming to the U.S. in 1935. In 1938 he developed a coffeemaker for his own use, derived from an ordinary Erlenmeyer chemical flask, funnel, and paper filter, which he patented, and in 1941 marketed his *Chemex* coffeemaker invention. It became an instant success and inspired the design of future coffeemakers. He developed a number of other products, none as successful as the *Chemex.*

Don A. Schreckengost (1911-2001) U.S. ceramics designer who taught ceramics design at Alfred University in the 1930s and freelanced for a number of ceramic companies. His 1934 *Tricorne Mandarin Orange* became a classic design. In 1942, he headed Homer Laughlin's design department after the death of Frederick Rhead, and was responsible for all its postwar patterns, including *Rythm, Epicure,* and *Skytone* series. Since 1965 he was head designer at Hall China. He was a Fellow of the American Ceramic Society and received the prestigious Binns Medal.

Viktor Schreckengost (b. 1906). U.S. industrial designer who was born in Sebring, Ohio, and studied cartooning at the Cleveland School of Art and at the Kunstegewerbschule in Austria. In 1930 he joined the faculty at the Cleveland School of Art and worked part-time at Cowan Pottery Studio. He worked for a variety of other industrial and ceramic companies, then began an industrial design program at the Cleveland Institute of Art in 1933, which became well known for its automotive design. He created stage and costume designs from 1938 to 1943, and served with the Navy in World War II. After the war, he designed many consumer products while continuing his work in fine art, with sculpture and painting. In 2001 he was honored with a retrospective exhibition with catalogue at the Cleveland Museum of Art, *Victor Schreckengost and 20th Century American Design.*

Joseph (Jo) Claude Sinel (1889-1975). Self-described as the first industrial designer in the U.S. because he used the title on his letterhead in 1920. Born in Auckland, New Zealand, he attended the Elam School of Art, and, after working in Australia and England, emigrated to San Francisco in 1918. He wrote *A Book of American Trademarks and Devices.* Working for the Lennon Mitchell advertising agency in New York, he began designing art deco products for clients in 1923. His package designs were featured in a 1934 *Fortune* magazine. He taught at the California College of Arts and Crafts in the 1920s and returned there in the 1940s. One of the 14 founders of the Society of Industrial Designers in 1944, he also taught at a number of other design schools including Pratt Institute in Brooklyn and Chouinard in Los Angeles.

David B. Smith (b. 1932). U.S. industrial designer who studied at Ohio State University and graduated from Pratt Institute in industrial design. He started his career as a staff designer for the Kelvinator Division of American Motors Corporation, later working for Chrysler and then AEG in Germany. In 1959, with Deane Richardson, he co-founded Richardson/Smith in Columbus Ohio. In 1989 Richardson/Smith merged with the London-based firm, Fitch & Company PLC. In 1994 he became the Nierenberg Distinguished Professor of Industrial Design at Carnegie Mellon University and received the Alumni Achievement Award from Pratt Institute. He served as Chairman, President, and Treasurer of the Association of Professional Design Firms, and was a member of the Human Factors Society. Over the years he has received additional awards, patents, recognition, and articles, including IDSA's Personal Recognition Award in 1995.

F. Eugene Smith U.S. industrial designer who graduated from the Cleveland Institute of Art in 1946. He spent a year with the George Walker office, then teamed with his classmate, Samue Scherr and Bernard McDermott to establish the Akron, Ohio firm of Smith, Scherr & McDermott in 1947. In 1960, Smith left the firm to establish his own office, F. Eugene Smith Associates, Inc and in 1966, lectured against the ugliness and visual chaos of urban environments. In 1977, he founded F. Eugene Smith/Design Management, Inc. in Akron, Ohio.

Robert G. Smith. (b.1922) U.S. industrial designer who, after five years in the U.S. Merchant Marine, re-entered Pratt Institute and graduated in 1951. He worked for Harley Earl Associates from 1954 to 1956, for Raymond Loewy Associates from 1956 to 1959, and from 1959 to 1967, was vice president of product development at Lippincott & Margulies. He joined J.C. Penney in 1967, where he organized and headed an internal product design staff, and in 1974 implemented an award-winning corporate identity program. He was president of IDSA 1981-1982 and remained with Penney until his retirement in 1983.

William Snaith (1908-1974) U.S. industrial designer who was born in Brooklyn, New York, studied architecture at New York University, and art at the Ecole des Beaux-Arts, and Fontainebleau Academy of Fine Arts in Paris. He worked as an architectural draftsman and set/costume designer, until offered a job by Raymond Loewy in 1936, leading to his partnership in the Loewy office specializing in marketing. He became president of Raymond Loewy/William Snaith in 1961 when Loewy became Chairman.

Ettore Sottsass Jr. (b. 1917) Leading post-war Italian architect and industrial designer who was born in Austria, studied at Turin Polytechnic University and began his architectural career in 1947. He worked for George Nelson briefly in 1956, then started the same year with Olivetti, where he designed a number of acclaimed products including the brilliant red *Valentine* typewriter in 1969. In 1980 he founded Sottsass Associati and in 1981 created an avant-garde furniture design group called *Memphis,* and designed its signature post-modern *Casablanca* sideboard. Memphis dissolved in 1988, and in 1997, he designed the Malpensa 2000 international airport in Milan.

Paul B. Specht (b. 1925). U.S. industrial designer who served in the U.S Army from 1942 to 1945, and graduated from the School of the Art Institute of Chicago in 1950. He operated his own firm, PBS Design, in Chicago, Illinois, from 1949 to 1952, and worked with Russel Wright Associates, New York City, as a staff designer from 1952 to 1953. In 1953, he joined Chapman, Goldsmith, Yamasaki Inc. in Chicago as executive vice-president, a position he held until 1970. In 1970 he was named as a partner in the firm that became known as Goldsmith Yamasaki Specht, Inc. He held that position until 1996. In 1998 he became the principal of PBSDESIGN, INC, and served as senior design counsel for SundbergFerar, Inc. until 2000. He holds over 35 utility patents. He served actively in the Industrial Designer's Society of America (IDSA) from 1953, including chair of the Chicago chapter in 1960-1962 and 1968-1970, and IDSA's national executive vice president from 1980 to 1982. He became a Fellow of the Society in 1957.

Raymond Spilman (1911-2000) U.S. industrial designer who was born in Wichita, Kansas, studied at Kansas State University 1933-1934, worked for General Motors as a stylist from 1935 to 1939, joined Walter Dorwin Teague Associates in 1940 as a staff designer, and also worked for Henry Dreyfuss Associates in New York. In 1942 he joined Johnson/Cushing/Nevell as director of product design before opening his own office in 1946 in Darien, Connecticut. He was president of ASID 1960-1962. An avid watercolorist, he retired to Cape Cod, Massachusetts around 1999.

Mart Stam (1899-1986) Dutch architect born Martinus Adrianus Stam who was associated with members of the Bauhaus and *Constructivists*; and was the inventor of the tubular steel cantilever chair concept first shown in 1927. Stam worked as an architect in the Soviet Union from 1930 to 1934 before taking up directorships of art academies in Amsterdam, Dresden, and East Berlin from 1939 until his retirement in 1966.

Budd Steinhilber (b. 1924). U.S. industrial designer who was born in New York City, grew up in Woodstock, New York, and studied at the High School of Music and Art in New York City. He apprenticed at the Loewy office in New York while still a student at Pratt Institute. He graduated in 1943 and started work with Dohner & Lippincott in New York. In 1947 he became part of the team that designed the Tucker car. In 1949 he joined Read Viemeister in Yellow Springs, Ohio, as a partner until 1964, when he relocated to San Francisco and established a partnership with Gene Tepper (Tepper & Steinhilber Associates). In 1975 he partnered with graphic designer Barry Deutsch, forming Steinhilber & Deutsch Inc. He taught at the Dayton Art Institute and lectured at many other schools. He relocated to Kailua-Kona, Hawaii in 1987, and served as Secretary Treasurer of IDSA from 1989 to 1990. He is a Fellow of IDSA. Since 1989 he served as a design advisor to Konawaena High School Solar Car Team, and in 1997 was an NGO delegate to the Global Warning Conference in Kyoto.

Varvarya Stepanova (1894-1958) Designer who was born in Lithuania, attended Arts School in Kazan (1910-1911), and attended Stroganov School in Moscow (1912-1914). She was noted for her avant-garde textile designs and sports clothes. She became a leading proponent of *Constructivism* as its co-founder with her husband, Aleksandr Rodchenko, and with the first working group of *Consructivists* in 1921. She taught in Russia at the,Fine Art Studio of the Academy of Social Education from 1921 to 1924, and on the textile faculty of the VKhUTEMAS from 1924 to 1925.

Clifford Brooks Stevens (1911-1995). U.S. industrial designer who was born in Milwaukee, Wisconsin, studied architecture at Cornell from 1929 to 1933, and immediately entered the profession by designing trademark materials for his father's firm, Cutler-Hammer. He worked on outboard motors for Evinrude starting in 1934, and established his own office. He re-designed the *Steam-O-Matic* steam iron in 1938, and in 1941 re-designed the Hiawatha Skytop Lounge for the Chicago, Milwaukee & St. Paul Railroad. He was one of the founders of SID in 1944. He became a consultant to Willys-Overland for post-war Jeep vehicles including the *Jeepster* in 1949 and the *Wagoneer* in 1963. He designed for Studebaker from 1962 to 1967, and designed the famous Oscar Mayer *Wienermobile*. After 1979, Brooks Stevens Design Associates continued under his son, Kipp. From 1983 until his death, Stevens taught at the Milwaukee Institute of Art & Design.

Gustav Stickley (1857-1942).A designer and follower of the Arts and Crafts movement who entered the furniture business with many brothers in 1886 and established his own firm in 1898, designing and making what became known as Mission furniture. In 1905 he moved to New York, but by 1915 was bankrupt, due to widespread imitation of his style by competitors. In 1884, Gustav's brother, **Charles Stickley,** formed the Stickley-Brandt Furniture Company in Binghamton, New York, and produced Victorian design furniture until it went bankrupt in 1919. Two brothers, **George and Albert Stickley**, moved to Grand Rapids, Michigan, in 1891, making furniture similar to Gustav's at their firm, Stickley Brothers Company, which closed in 1907. Two other brothers, **Leopold and J. George Stickley**, opened a factory in Fayetteville, New York, under the name L. and J. G. Stickley Company, and made designs similar to Gustav's, as well as some designed by Frank Lloyd Wright. In 1916, Leopold and George purchased Gustav's factory after his bankruptcy, and continued to operate it as the Stickley Manufacturing Company.

Gunta Stölzl (1897-1983) German, later Swiss designer who studied at the Kunstgewerbeschule in Munich (1914-1916) and at the Bauhaus (1919-1923). She joined Johannes Itten at Herrliberg in 1924 but returned to the Bauhaus in 1925 as a journeyman, then as Master (1926) and headed the weaving workshop from 1927 until her resignation in 1931. She was the only woman on the Bauhaus staff and in terms of designs sold to business, her department was the most successful at the Bauhaus. She established her own studio in Zurich with Bauhaus colleagues in 1931, and after 1937 ran the studio alone until 1967, meanwhile active in the *Schweizer Werkbund.*

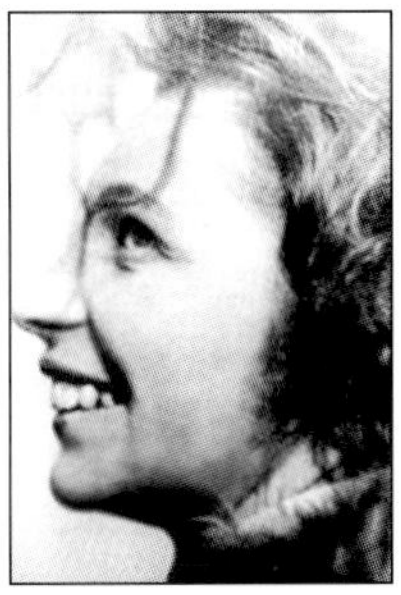

William Bushnell Stout (1880-1956) U.S. automotive and aeronautical designer who was born in Quincy, Illinois, and graduated from the University of Minnesota. He became the technical editor of the *Chicago Tribune* in 1912, worked for Packard in 1916, and in 1917 became an advisor to the National Aircraft Board in Washington, D.C. He founded the Stout Engineering Laboratories in 1919 and developed the first all-metal plane, the Ford *Tri-motor* in 1925. In 1932 he began to operate the Stout Research Division of Consolidated Vultee Aircraft Corporation. He developed a Pullman "Railplane" in 1934, the *Scarab* car in 1935, and four airplane/automobile "Skycar" concepts from 1931 through 1945.

William Stumpf. (b. 1936) U.S. industrial designer who received a degree in industrial design from the University of Illinois in 1959, and a masters in environmental design from the University of Wisconsin in 1968. In 1970 he joined the Herman Miller Furniture Company as director of research. In 1973 he established a consulting practice, Stumpf Associates, in Winona, Minnesota and designed the 1976 *Ergon* chair for Herman Miller. He joined with Don Chadwick in 1977 to form Chadwick Stumpf and Associates and the result was the 1984 *Equa* chair and Ethospace System, both for Herman Miller.

Louis Henri Sullivan (1856-1915) U.S. architect who grew up in Boston, studied architecture at MIT, and moved to Chicago in 1873. He worked for architect William Jenney and in 1874 left to study at École des Beaux-Arts, Paris. He joined the firm of Dankmar Adler (1844-1900) in Chicago as junior partner in 1882, became a full partner in 1883, and left in 1895 for his own practice. He became well known for his nature-related and *Art Nouveau* style, and for early skyscrapers. In 1888 he employed Frank Lloyd Wright who called him "Lieber Meister". Sullivan frequently used the phrase often credited to him, "Form ever follows Function."

Carl W. Sundberg (1910-1982). U.S. industrial designer who was born in Calumet, Mississippi, graduated from Northern High School and attended Wicker Art School in Detroit. He began his career in 1927 designing custom car bodies under Ray Dietrich at Dietrich Body Company. He then worked in plastics design at Kurz-Kasch, Inc. and then later at General Motors (GM) where he met Montgomery Ferar. In 1934, he and Ferar left GM and formed the lifelong partnership of Sundberg-Ferar, Inc., designing major transportation systems and major appliances for Sears Roebuck and Whirlpool. He retired in 1975.

Anne Swainson (c.1900-1955) Swedish designer who emigrated to the U.S. and studied Fine and Applied Arts at Columbia University. She later taught Textile and Applied Design at Berkeley, worked as a fashion coordinator at Chase Revere Copper & Brass, and then founded and headed the Bureau of Design at Montgomery Ward in 1931.

Edward Russell Swann (1905-1958) U.S. industrial designer who was born in Jewett, Ohio, raised in Canton, Ohio, attended Akron University and joined the Hoover Company in North Canton, Ohio in 1925. In 1942 he was sent by Hoover for four years to work in the office of Henry Dreyfuss, Hoover's design consultant, learning the skills of industrial design. He returned to Hoover in 1946 to become the lead industrial designer on staff, while Dreyfuss continued to serve as Hoover's industrial design consultant. Swann became an early member of SID and continued to head the Hoover industrial design department until his death.

Vladimir Tatlin (1885-1953) Leading theorist of Russian *Constructivism* who studied painting at the Moscow School of Painting, Sculpture and Architecture (1902-1904) and at Penza Art School (1904-1910) and from 1910 to 1920 was linked to Russian *Futurism.* After the 1917 revolution he became responsible for Lenin's Plan for Monumental Propaganda. He moved to Petrograd in 1919 where he worked on a model for his famous *Monument to the Third International,* exhibited in 1920. From 1927 on, he taught industrial design (called 'production art') at the VKhUTEMAS.

Richard Teague (1923-1991) U.S. automotive designer who was born in Los Angeles and studied at the University of Southern California and the Art Center College of Design. He worked for Henry Kaiser on the design of a small economy car and in 1948 joined General Motors where he worked on the 1949 Oldsmobile *Rocket 88* in the Cadillac studio. In 1952 he joined the Packard Motor Car Company as chief stylist, and restyled the 1953 *Caribbean.* In 1959 he began his long career at American Motors, becoming styling director in 1962 and design vice-president in 1964. He retired in 1984.

Walter Dorwin Teague (1883-1960). U.S. industrial designer who was born in Decatur, Indiana, and is considered by many to be the dean of industrial design. He studied at the Art Students League in New York from 1903 to 1907. He worked as an illustrator, establishing his own commercial art studio in 1911. By the mid-1920s, he was well known for his elegant advertising drawings. He left advertising to open one of the first industrial design firms in 1927, receiving his first contract with Eastman Kodak in 1928. He was on the Board of Design for the 1939 World's Fair, where he also designed the Ford and U.S. Steel pavilions. In 1944, he was one of the founders of the Society of Industrial Designers, becoming its first president. After World War II, he became a major consultant for Boeing, designing many interiors of the jet age.

John J. "Jack" Telnack (b. 1937) U.S. automotive designer who graduated from Art Center College in 1958 and joined Ford. He worked on the fastback version of the *Mustang* in the late 1960s. He was transferred to Australia, then to Ford of Europe, where he worked on the 1977 *Fiesta*. He returned to the U.S. and headed Ford's North American design studio, focused on the design of the aero-look 1983 *Thunderbird* and then on the 1985 *Taurus* and *Sable*. He became vice president of corporate design at Ford in 1987, a position he held until his retirement in 1997.

Louis Comfort Tiffany (1848-1933). U.S. artist and designer who was the son of Charles Lewis Tiffany (1812-1902), founder of Tiffany Jewelry and Silversmiths in New York. He was best known for his work in glass and metal with a strong Art Nouveau flavor. He traveled to Europe in 1865, and returned to study painting with George Inness. He established his own studios in 1879 as Louis C. Tiffany & Associated Artists, and established the Tiffany Glass Company in 1886. He later inherited his father's separate business in New York.

John Tjaarda ("charda") (1897-1962) U.S. product and automotive designer who was born in Holland, trained in aeronautical design in England and served as a Dutch Air Force pilot. He emigrated to the U.S. in 1923 and worked on custom car bodies in Hollywood, then pioneered on monocoque streamlined designs. He worked for Duesenberg from 1928 to 1930, and for Harley Earl from 1930 until 1932, when he became chief of body design for the Briggs Manufacturing Company. His streamlined concept *Brigg's Dream Car* for Ford displayed at the *Century of Progress Exhibition* in 1933 led to the development of the 1936 Lincoln *Zephyr*.

David Tompkins (b. 1934) U.S. industrial designer who was born in Glen Gardner, New Jersey, and received an industrial design degree from Pratt Institute in 1961. In 1963 he joined the new firm of Richardson/Smith in Worthington, Ohio where he worked on a range of projects. He left Richardson/Smith in 1977 to open his own office, Tompkins & Associates, and in 1981 became vice president of design for RCA Consumer Electronics in Indianapolis, Indiana. In 1987 he joined Montell Polyolefins to establish a design/engineering center in Wilmington, Delaware. He is a Fellow and past president of IDSA (1993-1994) and retired in 1995 to Florida.

Alex Sarantos Tremulis (1914-1994) U.S. automotive designer who was born in Chicago, started at Duesenberg in 1933, and became chief stylist in 1936. In 1937 he joined General Motors, then Briggs-LeBaron. In 1939 he consulted with Crosley and American Bantam but returned to Briggs to work on the 1941 Chrysler *Thunderbolt* and Packard *Clipper*. He entered the Air Force in 1941, where he worked on advanced aircraft concepts. After the war, he worked with Tammen and Denison until Preston Tucker hired him in 1947 to work on the *Tucker 48*. He later worked at Ford advanced styling, and in California for Subaru, where he designed the *Brat*.

Joseph Urban (1872-1933) Austrian architect who studied architecture at the Royal Academy and became a member if the *Vienna Secession* of 1897. By 1908 he was an illustrator and designer of theatrical sets. He emigrated to the U.S. in 1911 to become artistic director of the Boston Opera. He moved to New York and became famous as set designer for the Zeigfield Follies from 1918 to 1933. He operated a New York branch of the *Wiener Werkstätte* from 1922 to 1924, and opened his own office in New York in 1925 where he practiced interior design, designed typical *Art Deco* furniture, and continued work as set designer.

Henry Van de Velde (1863-1957). Belgian painter-designer-architect and one of a group known as *Les Vingt* (The Twenty), who absorbed the principles of the English Arts and Crafts movement and the French Symbolists, and transformed them into the *Art Nouveau* style. He founded his own firm and factory near Brussels c. 1899, and his Art Nouveau furniture designs were well known. In 1902 he founded the *Weimar Kunstgewerbeschule* (Weimar School of Arts and Crafts), the precursor of the Bauhaus, which became public in 1907. He was a founder member of the *Deutsche Werkbund,* but had a disagreement in 1914 with Hermann Muthesius, who believed in the standardization of design. Van de Velde later resigned his post, recommending Walter Gropius as his successor. In 1917 he moved to Switzerland, in 1920 to the Netherlands, and in 1947 returned to Switzerland.

Theo van Doesburg (1893-1931). Dutch artist born in Utrecht, the Netherlands, whose given name was C.E.M. Kupper. He began as a painter but was drawn to architecture in 1916 by J.J.P. Oud. He was the founder and leader of the *de Stijl* movement, disseminating its ideas through lecture tours and its magazine, *de Stijl,* published from 1917 until his death. He taught briefly at the Bauhaus in 1922.

Harold Van Doren (1895-1957). U.S industrial designer who was born in Chicago, Illinois, and graduated from Williams College in 1917. He studied at the Art Students League in New York (1920-21), and in Paris (1922-24). In 1927 he became assistant director of the Minneapolis Institute of Arts. He opened a design office in 1931 with John Gordon Rideout in Toledo, Ohio. In 1935, Rideout left and the firm became Harold Van Doren & Associates, which relocated to Philadelphia in 1940, the same year he wrote the book, *Industrial Design: A Practical Guide.* In 1944 he became a co-founder of SID, and his firm was expanded to include a New York office, known as Van Doren, Nowland & Schladermundt. In 1950, the partners separated, with Van Doren retaining his office in Philadelphia. Upon his death, the firm became known as Harper Landell & Associates, and later, Bresslergroup, Inc.

John Vassos (1898-1985). U.S. industrial designer who was born in Constantinople (now Istanbul, Turkey), the son of a Greek publisher. While serving on a suicide minesweeper in World War I, he was torpedoed, rescued, and landed in Boston in 1919. He studied art and illustration with John Singer Sargent at the Fenway Art School. He came to New York in 1924 where he studied at the Art Students League and opened an industrial design studio. He published and illustrated a number of books from 1929 to 1931. In 1933 he established the first internal design department at RCA, remaining its consultant until 1964. He was the first president of the American Design Institute (ADI) in 1938. During World War II, with a predecessor of the CIA, he helped organize the Greek underground and developed camouflage techniques. He was elected the first chairman of the board of IDSA when it was formed in 1965.

Read Viemeister (1923-1993). U.S. industrial designer who graduated from Pratt Institute in 1943 along with Budd Steinhilber; both started work with Dohner & Lippincott in New York, with Viemeister becoming its first director of styling. He married Beverly Lipsett in 1946, moved to Yellow Springs, Ohio, and organized Vie Design Studios with Steinhilber as partner in 1947. Both were on the team that designed the 1948 Tucker automobile. In 1949 they worked together to establish a department of industrial design at the Dayton Art Institute, where they both taught and lectured. When Steinhilber left for San Francisco in 1964, Viermeister maintained Vie Design with the help of Frank Burrer and others until the 1990s.

Tucker Viemeister (b. 1948) U.S. industrial designer and son of Read Viemeister, was born in Yellow Springs, Ohio, received a BFA from Pratt Institute in 1974, then worked in his father's industrial and graphic design firm, Vie Design Studios. In 1979 he co-founded Smart Design Inc. in New York City with Davin Stowell, Tom Dair and Tamara Thomsen. Their best -known designs are *GoodGrips* kitchen tools for Oxo International. In 1997 he headed *frogdesign's* New York studio, in 1999, he became executive vice president of the Research and Development Department at Razorfish, and recently, became president of Springtime-USA.

Sohrab Vossoughi. U.S. Industrial designer who was born in Tehran, Iran in 1956, and moved to the United States in 1971. After studying mechanical engineering for three years, he switched to study industrial design, graduated from San Jose State University's Department of Industrial Design in 1979, and joined Hewlett-Packard Corporation. In 1982, he began independent consulting for startup companies in Portland, Oregon and by 1984, had launched a product development firm he named ZIBA Design. In 1991 a Ziba office was established in Tokyo. By 1994, Ziba had a staff of 28, and had won more design awards than any other US design firm.

Wilhelm Wagenfeld (1900-1990). Craftsman and educator associated with the Bauhaus, who was born in Germany, trained as goldsmith, and studied drawing at Hanau, before becoming a Bauhaus student in 1923 in the metal workshops under László Moholy-Nagy. He left the Bauhaus in 1929 and worked in porcelain and glass with Lausitzer Glasverein (1935-1938) and Jenaer Glaswerke. He taught design at the Berlin Hochshule für Bildende Künste. He designed an unusual 1938 *Zigzag* Pelican inkbottle that is still in production, and in 1954 established a workshop in Stuttgart. He was a co-founder of the New Werkbund in 1947 and of the journal *Form* in 1957.

Otto Wagner (1841-1918). Austrian architect, designer, and city planner who was a leading force in the *Wiener Sezession* (Vienna Secession) formed in 1897. Considered by many to be the world's first modernist, he was trained at the Vienna Polytechnic and the Bauakademie in Berlin, and in 1894 became the leading professor at the Akademie der bildenden Künst in Vienna. In 1895 he moved to a form of early modernism, and in 1900 became the city architect of Vienna. His most important work is considered to be the Vienna Postal Savings Bank (1904-1912).

George W. Walker (1908-1993) U.S. automotive designer who trained at Otis Institute, Los Angeles in 1916, continued at Cleveland School of Art, and headed his own industrial design office in Detroit by the 1930s. His firm worked for Nash (1937-1945) and in product design as well. His role at Ford began in 1945 as an independent consultant working on the 1949 Ford, Ford's first true post-war model. He became Ford's first vice president of styling in 1955, where he remained until his retirement in 1962.

Don Wallance (1909-1990). U.S. furniture and metalware designer who received a BA from New York University and trained at the Design Laboratory, New York (1935-1939). In 1938, the Museum of Modern Art awarded him a prize for a chair he designed for its new building. During World War II he researched mass-produced furniture for the Armed Forces and conducted a study of plywood storage units that won a prize at Museum of Modern Art's low-cost furniture competition in 1948. He designed Lauffer's *Design 1 and* other stainless steel flatware patterns. He designed contract seating for Philharmonic Hall, Lincoln Center, and wrote *Shaping America's Products* in 1956.

Arnold Wasserman U.S. industrial designer who was born in Chambersburg, Pennsylvania and graduated from Carnegie Institute of Technology in 1956. In 1975 he formed Tag, Inc. to produce the soft luggage he had designed. In 1980, as director of corporate industrial design at the National Cash Register Corporation, he implemented comprehensive corporate design strategy and standards. He later held similar corporate design strategy positions at Xerox and Unisys. In 2003 he became chairman of the Idea Factory in Singapore, redesigning the educational system, media and design industries there. He teaches an interdisciplinary course at the University of California, Berkeley's Haas Graduate School of Business.

Kem Weber (1889-1963) German industrial designer who apprenticed as a cabinetmaker and graduated from the Kunstgewerbemuseum in Berlin in 1912. In 1914 he was sent as a designer for Germany's exhibit at the Panama -Pacific Exposition in San Francisco, was stranded there by World War I, and settled in the U.S. From 1922-1927 he was art director for the Barker Brothers store in Los Angeles where he introduced his modern style to the shop he designed there. After 1927, he worked for numerous other manufacturers. His best known work was his 1935 *Airline* chair. Toward the end of the 1930s he worked mostly on interiors and architecture.

Hans Jorgensen Wegner (b.1914). Danish furniture designer who started apprenticeship as a cabinetmaker at the age of 14, and completed it in 1931. He trained as a furniture designer at the School of Arts and Crafts in Copenhagen, and by 1938 was in the office of Arne Jacobsen and Eric Moller. In 1940 he began an association with the Copenhagen cabinetmaker Johannes Hansen. His first important chair was the *Peacock* chair (1947), and his most important was the *Round* chair (1949).

Frank Lloyd Wright (1867-1959). Widely regarded as the most significant architect of the 20th century, and certainly the best known, he was born in Wisconsin and studied engineering at the University of Wisconsin (1885-1887). In 1888 he joined the architectural office of Adler and Sullivan, but left in 1893 to establish a partnership with a co-worker, Cecil Corwin, and designed the Winslow House. He opened his own office in 1896, and soon perfected his "Prairie Style" in the Roberts House (1904) and Robie House (1909). In 1911 his work was published in Europe in the *Wasmuth Edition*, by which time he had adopted his signature costume, emulating Elbert Hubbard. During 1915-1922 he was in Japan designing the Imperial Hotel in Tokyo, which survived an earthquake in 1923. In 1936 he completed *Fallingwater*, his most famous work, and the S.C. Johnson building in Racine, Wisconsin (1936-1939). His Guggenheim Museum in New York was completed in 1959.

Russel Wright (1904-1976). U.S. industrial designer who, between 1921 and 1924, attended the Art Students League in New York and the Columbia University of Architecture. He began his career in the theater as an apprentice with Norman Bel Geddes, then as a stage designer. He began a business with his wife Mary in 1930, designing and producing spun-aluminum accessories for the home. He designed *American Modern* furniture for Macy's in 1935, and, in 1939, *American Modern* ceramic dinnerware. He retired to *Dragon Rock*, his 79-acre estate on the Hudson River, in 1941. He was one of the founders of SID in 1944, and championed informal postwar living. His 1953 melamine plastic *Residential* table service was one of the earliest designs in plastic. He closed his office in 1958 and toured Asia developing cottage industries for the Department of State. In 1967 he began consulting with the National Park Service.

Marco Zanuso (b. 1916) Italian designer who studied architecture at Politecnico in Milan, where he became a teacher and director of the architectural faculty. He established his practice at the end of World War II and served as editor of *Domus* and *Casabella* magazines, as well as designer of a number of *Triennales* between 1947 and 1955. His Borletti sewing machine won a *Compasso d'Oro* in 1956. He began collaboration with Richard Sapper in 1958 with whom he designed a range of BrionVega television products. He has also designed industrial plants and offices for Olivetti.

Eva Zeisel (b.1906). Ceramics designer who was born Eva Striker in Budapest, Hungary, and trained at the Képzomuveszeti Academia (Academy of Fine Arts) in Budapest. She opened a studio in Hungary in 1925, and worked in Germany from 1928 to 1931. From 1932 to 1937 she worked in Leningrad and Moscow, becoming art director of the china and glass industry of the Russian Republic in 1935. She went to England in 1938, where she married Hans Zeisel and emigrated to New York with him. She started teaching in 1939 at Pratt Institute, and, after World Wart II, won a competition by the Museum of Modern Art to design *Museum Service* dinnerware. She worked for Red Wing Potteries, Italcraft, and Hall China, taught at Pratt until 1953, and at RISD until the 1960s. In 1982, she returned to Hungary to work in a ceramics factory. Back in the U.S., she was honored by exhibitions of her work mounted by the Brooklyn Museum, and another at the Museé des Arts Décoratifs in Montreal in 1984.

Photograph Credits

Chapter 1

01-01 Courtesy of © Eastman Kodak Company • **01-02** Photograph courtesy of Thonet Industries **01-03** Photograph courtesy of Thonet Industries • **01-04** Photograph courtesy of Gibson Guitar Corp., Nashville, TN • **01-05** Frank Lloyd Wright, designer. Dining room chair for oak Park Home, c. 1895, oak frame with leather seat, 57 1/2 X 18 X 19 1/2 in., Collection of Frank Lloyd Wright Preservation Trust, Gift of Olgivanna Lloyd Wright, 1977.06. Photographer: Philip Mrozinski. The Frank Lloyd Wright Preservation Trust • **01-06** Property of AT&T Archives. Reprinted with permission of AT&T • **01-07** Courtesy of Thomson Consumer Electronics • **01-08** Photograph courtesy of Thomson Consumer Electronics • **01-09** General Motors Corporation. Used with permission, GM Media Archives • **01-10** Property of AT&T Archives. Reprinted with permission of AT&T • **01-11** Mackintosh, Charles Rennie. Ladderback chair. (1902) Ebonized wood, velvet upholstery, 55 1/4 X 16 X 13 1/2" (140.4 X 40.7 X 34.3 cm). The Museum of Modern Art, New York. Estée and Joseph Lauder Design Fund. Photograph © 1997 The Museum of Modern Art, New York. Digital Image © The Museum of Modern Art/Licensed by SCALA/Art Resource, NY • **01-12** From the Collection of The Henry Ford • **01-13** Property of AT&T Archives. Reprinted with permission of AT&T • **01-15** Photograph courtesy of the author • **01-16** Photograph courtesy of The Gillette Company • **01-17** From the Collection of The Henry Ford • **01-18** National Air and Space Museum, Smithsonian Institution (SI 2002-16646) • **01-19** Binney & Smith properties, Inc., Easton, Pennsylvania • **01-21** Riemerschmid, Richard. Armchair. (1899) Oak, leather, 32 13/16 X 20 7/8 X 20 7/8" (83.1 X 53 X 53.3 cm). The Museum of Modern Art, New York. Joseph H. Heil Fund. Photograph © 1997 The Museum of Modern Art, New York. Digital Image © The Museum of Modern Art/Licensed by SCALA/Art Resource, NY • **01-22** Virginia Museum of Fine Arts, Richmond. Gift of the Sydney and Francis Lewis Foundation, photo: Katherine Wetzel, © Virginia Museum of Fine Arts • **01-23** Courtesy The Frank Lloyd Wright Archives, Scottsdale, Arizona • **01-24** Missouri Historical Society, St. Louis • **01-25** Los Angeles County Museum of Art, purchased with funds provided by Elma Shoemaker, Daisy and Daniel Belin, and Susan and Robert F. Maguire, III, through the 1989 Collectors Committee. Photograph © 2004 Museum Associates/LACMA • **01-26** Property of AT&T Archives. Reprinted with permission of AT&T • **01-27** Schenectady Museum • **01-28** The NCR Archive at the Montgomery County Historical Society • **01-29** Photograph courtesy of Carnegie Library of Pittsburgh • **01-30** Photograph courtesy of Thomson Consumer Electronics • **01-31** Photograph courtesy of Bissell Homecare Division • **01-32** Los Angeles County Museum of Art, Gift of Max Palevsky. Photograph © 2004 Museum Associates/LACMA • **01-34** Photograph courtesy of the Detroit Public Library, National Automotive History Collection • **01-35** Deutsches Technikmuseum Berlin, AEG-Archiv • **01-36** Courtesy of Maytag • **01-37** Property of AT&T Archives. Reprinted with permission of AT&T • **01-38** Photograph courtesy of Vitra Design Museum, Weil am Rhein • **01-39** Deutsches Technikmuseum Berlin, AEG-Archiv • **01-40** Photograph courtesy of the International Housewares Association • **01-41** Deutsches Technikmuseum Berlin, AEG-Archiv • **01-42** Schenectady Museum • **01-43** Deutsches Technikmuseum Berlin, AEG-Archiv • **01-44** Deutsches Technikmuseum Berlin, AEG-Archiv • **01-45** Los Angeles County Museum of Art, gift of Max Palevsky in honor of the museum's twenty-fifth anniversary. Photograph © 2004 Museum Associates/LACMA • **01-46** Gustav Stickley, American, 1857-1942, Settle, 1909, oak, , back: 38 x 71 1/16 x 22 in., seat: 19 x 62 in., Gift of Mr. and Mrs. John J. Evans, Jr., 1971.748. Reproduction, The Art Institute of Chicago.

Chapter 2

02-01 Schenectady Museum • **02-02** Hagley Museum and Library, Wilmington, Delaware • **02-04** Photograph courtesy of Stiftung Bauhaus, Dessau and Greten GmbH & Co. • **02-05** U.S. Department of the Interior, National Park Service, Edison National Historic Site • **02-06** Photograph courtesy of the Detroit Public Library, National Automotive History Collection • **02-08** From the collection of Roycroft Arts Museum, Boice Lydell, East Aurora, New York 1912 catalog, p. 3. Photograph by the author • **02-09** American Aviation Historical Society, photograph courtesy of John Sprogge • **02-10** Courtesy of the International Housewares Association • **02-11** Photograph courtesy of Friday Historical Business Archives, Louisville, Kentucky • **02-12** From the Collection of The Henry Ford • **02-13** Courtesy of the International Housewares Association • **02-14** Copyright DaimlerChrysler Corporation. Used with permission • **02-15** Photograph courtesy of the Detroit Public Library, National Automotive History Collection • **02-16** Photograph courtesy of Wallace H. Appel, FIDSA • **02-17** Coca-Cola is a trademark of The Coca-Cola Company • **02-18** Collection of The Corning Museum of Glass, Corning, NY, gift of Jerry E. Wright • **02-19** Photograph courtesy of the author • **02-20** Hugh Moore Dixie Cup Collection, David Bishop Skillman Library, Lafayette College, Easton, Pennsylvania • **02-21** Photograph courtesy of the Detroit Public Library, National Automotive History Collection • **02-23** Photograph courtesy of Black & Decker • **02-24** National Air and Space Museum, Smithsonian Institution (SI 92-14959) • **02-25** Photograph courtesy of the author • **02-26** Rietveld, Gerrit (1888-1964) © ARS, NY. Red and Blue Chair, c. 1918. Wood, painted, 34 1/8 X 26 X 26 1/2". Seat height: 13". Gift of Philip Johnson. The Museum of Modern Art, New York, U.S.A.. Digital Image © The Museum of Modern Art/Licensed by SCALA/Art Resource, NY • **02-27** Photograph courtesy of Douglas F. Wolff, IDSA • **02-28** Photograph courtesy of the author • **02-29** Photograph used with permission of KitchenAid, courtesy of Whirlpool Corporation • **02-30** Property of AT&T Archives. Reprinted with permission of AT&T • **02-31** Photograph courtesy of Automobile Quarterly.

Chapter 3

03-01 Tatlin, Vladimir. Model of the Monument to the Third International. 1920. Photograph courtesy, The Museum of Modern Art, New York. Digital Image © The Museum of Modern Art/Licensed by SCALA/Art Resource, NY • **03-02** Photograph courtesy of Carnegie Library of Pittsburgh • **03-03** Image courtesy of the author • **03-04** General Motors Corporation. Used with permission, GM Media Archives • **03-05** Deutsches Technikmuseum Berlin, AEG-Archiv • **03-06** Image courtesy of the author • **03-07** Hugh Ferriss, American, 1889-1962. Study for Maximum Mass Permitted by the 1916 New York Zoning Law, Stage 2, 1922. Black crayon, stumped, brush, and black ink, over photostat, varnish, on illustration board 25 5/16 X 20 1/16 inches (668 X 509 mm). Cooper -Hewitt, National Design Museum, Smithsonian Institution. Gift of Mrs. Hugh Ferriss 1969-137-2 • **03-08** Schenectady Museum • **03-09** Photograph courtesy of Automobile Quarterly • **03-10** Wagenfeld, Wilhelm (1900-1990). Table lamp. 1923-24. Glass and chrome-plated metal. 18 X 8" (45.7 X 20.3 cm); diam.5 1/2" (14 cm). Gift of Philip Johnson. (490.1953) The Museum of Modern Art, New York. NY, U.S.A. Digital Image © The Museum of Modern Art/Licensed by SCALA/Art Resource, NY • **03-11** Photograph by Staatliche Bildstelle, Berlin, 1923, provided courtesy of Bauhaus Archive, Berlin • **03-12** Getty Images • **03-13** Copyright DaimlerChrysler Corporation. Used with permission • **03-14** Photograph courtesy of Stiftung Bauhaus Dessau • **03-15** Photograph courtesy of the International Housewares Association and Electrolux Home Care Products, North America • **03-16** Le Corbusier and Jeanneret. Pavillion de L'Esprit Nouveau, International Exposition of Decorative Arts. Interior, Living Room, Paris, 1925. Photograph courtesy, The Museum of Modern Art, New York. Digital Image © The Museum of Modern Art/Licensed by SCALA/Art Resource, NY • **03-17** Gropius, Walter. Bauhaus Building. Dessau, Germany 1925-26. Photograph courtesy, the Museum of Modern Art, New York. Digital Image © The Museum of Modern Art/Licensed by SCALA/Art Resource, NY • **03-18** From the Collection of The Henry Ford • **03-19** National Museum of American History, Smithsonian Institution, Behring Center • **03-20** General Motors Corporation. Used with permission, GM Media Archives • **03-21** Photograph courtesy of the Detroit Public Library, National Automotive History Collection • **03-22** National Air and Space Museum (NASM A-48743-W),Smithsonian Institution • **03-23** Photograph courtesy of Knoll, Inc. • **03-24** Photograph courtesy of Knoll, Inc. • **03-25** Schenectady Museum • **03-27** Collection of Cranbrook Art Museum • **03-28** Photograph by Peter Kühn, courtesy of Stiftung Bauhaus Dessau and Peter Kühn • **03-29** Photograph by Peter Kühn, courtesy of Stiftung Bauhaus Dessau and Peter Kühn • **03-30** Mies van der Rohe, Ludwig (1886-1969). © ARS, NY. "Barcelona" Chair. 1929. Stainless steel bars; leather-upholstered cushions, height 29 3/8"; width, 29 1/4"; depth, 29 3/4"; Seat height, 17 3/8" . Manufacturer: Knoll International, New York, New York, U.S.A. Gift of Knoll International (552.1953). The Museum of Modern Art, New York, New York, U.S.A. Digital Image © The Museum of Modern Art/Licensed by SCALA/Art Resource, NY • **03-31** Photograph courtesy of Knoll, Inc. • **03-32** National Air and Space Museum, Smithsonian Institution (SI 72-201) • **03-33** Photograph courtesy of the International Housewares Association.

Chapter 4

04-01 carbon copy of Petition, Specifications, and Photostat for Patent #2147, Westchester County, New York for "new and original CORNER SINK with SINGLE DRAINBOARD" designed by Gustav B. Jensen for th firm of Byrnes, Stebbins, parmelee & Blenko of Pittsburgh, Pa. dated December 30, 1930. Gift of Mrs. Ellen Gustav Jensen, 1956-47-50, Cooper-Hewitt, National Design Museum, Smithsonian Institution • **04-02** Photograph courtesy of Knoll, Inc • **04-03** Courtesy Miele & CIE. GmbH & Co. • **04-04** Photograph courtesy of the author • **04-05** Photograph courtesy of Knoll, Inc. • **04-06** National Air and Space Museum, Smithsonian Institution (SI 84-18230) • **04-07** Image courtesy of the Viktor Schreckengost Foundation • **04-08** Installation view of "Modern Architecture-International Exhibition." The Museum of Modern Art, New York. February 10. 1932 through March 23, 1932. Photograph © 1995 The Museum of Modern Art, New York. Digital Image © The Museum of Modern Art/Licensed by SCALA/Art Resource, NY • **04-09** Walter Dorwin Teague Associates, Inc. • **04-11** Schenectady Museum • **04-12** Syracuse University Library, Special Collections • **04-13** Smithsonian Institution, Transportation • **04-14** Photograph courtesy of the Detroit Public Library, National Automotive History Collection • **04-15** Installation view of the exhibition "Machine Art." The Museum of Modern Art, New York. March 5, 1934 through April 29, 1934. Photograph by Wurts Brothers. Courtesy The Museum of Modern Art, New York, NY, U.S.A. Digital Image © The Museum of Modern Art/ Licensed by SCALA/Art Resource, NY • **04-16** Photograph courtesy of Rubbermaid Home Products • **04-17** Smithsonian Institution, Transportation • **04-18** Photo courtesy of the Industrial Designers Society of America. Copyright DaimlerChrysler Corporation. Used with permission • **04-19** Courtesy of © Eastman Kodak Company • **04-20** National Air and Space Museum, Smithsonian Institution (SI 95-8328) • **04-21** From the Collection of The Henry Ford • **04-22** Syracuse University Library, Special Collections • **04-23** Courtesy of the Hoover Historical Center **04-24** From the Collection of The Henry Ford • **04-26** Courtesy of Erika and Robert Schott • **04-27** Property of AT&T Archives. Reprinted with permission of AT&T • **04-28** Photograph used with permission of KitchenAid, courtesy of Whirlpool Corporation • **04-29** Courtesy of Paul R. Wiedman, Peter Muller Munk Associates • **04-30** Hagley Museum & Library, Wilmington, Delaware • **04-31** Group of chrome toasters. David Painter Collection, Charles McKinney, Chicago, photographer, Courtesy of The Art Institute of Chicago. • **04-32** Courtesy of Volkswagen AG • **04-33** Syracuse University Library, Special Collections • **04-34** Photograph courtesy of the Industrial Designers Society of America • **04-35** From the Collection of The Henry Ford • **04-36** Photograph courtesy of the Industrial Designers Society of America.

Chapter 5

05-01 From the Collection of The Henry Ford • **05-02** 3-M tape dispenser, designed by Barnes & Reinecke. David Painter Collection. Shigeta-Wright, Inc., photographer. Courtesy of The Art Institute of Chicago • **05-04** Brooks Stevens *Steam-O-Matic Iron*, 1940, Photograph, Milwaukee Art Museum, Brooks Stevens Archive • **05-05** Schlumbohm, Peter. "Chemex" coffee maker. (1941) Pyrex glass, wood, leather, 9 1/2" h. X 6 1/8" diam. at rim X 2 3/4" diam. at waist, (24.2 X 15.5 X 7 cm) (including wood). The Museum of Modern Art, New York. Gift of Lewis & Conger. Photograph © 1997 The Museum of Modern Art, New York. Digital Image © The Museum of Modern Art/Licensed by SCALA/Art Resource, NY • **05-06** Copyright DaimlerChrysler Corporation. Used with permission • **05-08** Copyright DaimlerChrysler Corporation. Used with permission • **05-09** Photograph courtesy of the Detroit Public Library, National Automotive History Collection • **05-10** Courtesy of the Milwaukee Institute of Art & Design • **05-11** American Aviation Historical Society from Lockheed Photo • **05-12** Getty Images • **05-13** National Air and Space Museum, Smithsonian Institution.TM & © Boeing. Used under license • **05-14** Reproduced with Permission from Motorola Archives. © 2004, Motorola, Inc. • **05-15** Photograph courtesy of Knoll, Inc. • **05-16** National Air and Space Museum, Smithsonian Institution (SI 96-15926) • **05-17** National Air and Space Museum, Smithsonian Institution (SI 96-15924) • **05-18** Photo of Brunswick Bowling/Recreation Center Ball Return, Donald Deskey Collection, Cooper-Hewitt, National Design Museum, Smithsonian Institution • **05-19** Getty Images • **05-20** Photograph courtesy of Emeco • **05-21** Zeisel, Eva. Dinnerware (c. 1946). Porcelain, glazed; small oval dish: 1 1/2 X 5 1/16 X 2 9/16" (3.8 X 12.9 X 6.5 cm); creamer: 2 1/2 X 4 1/4" diam. (6.3 X 10.8 cm); divided relish dish with two spoons, 1 1/2 X 11 3/8 X 6 1/2" (3.8 X 28.9 X 16.5 cm); 7 5/16 X 1 11/16 " (18.6 X 4.3 cm) (each). The Museum of Modern Art, New York. Gift of the designer. Photograph © 1997 The Museum of Modern Art, New York. Digital Image © The Museum of Modern Art/Licensed by SCALA/Art Resource, NY • **05-22** Photograph courtesy of Tupperware • **05-23** Photograph courtesy of the International Housewares Association • **05-24** Photograph courtesy of Gibson Guitar Corp., Nashville, TN • **05-25** Photograph courtesy of the International Housewares Association • **05-27** Photograph courtesy of the author • **05-28** Coca-Cola is a trademark of The Coca-Cola Company • **05-29** Photograph courtesy of Tucker Madawick, FIDSA • **05-30** Photograph by Bill Sharpe, courtesy of Herman Miller, Inc. • **05-31** Photograph courtesy of Knoll, Inc. • **05-32** Photograph courtesy of the Detroit Public Library, National Automotive History Collection • **05-33** Courtesy of Polaroid Corporate Archives • **05-34** From the Collection of The Henry Ford • **05-35** Property of AT&T Archives. Reprinted with permission of AT&T • **05-36** Photograph by Rooks Photography, courtesy of Herman Miller, Inc.

Chapter 6

06-01 Photograph courtesy of the Industrial Designers Society of America • **06-02** Photograph courtesy of the Detroit Public Library, National Automotive History Collection • **06-03** Photo of Procter & Gamble *Crest* packaging, Donald Deskey Collection, Cooper -Hewitt, National Design Museum, Smithsonian Institution • **06-04** Syracuse University Library, Special Collections • **06-05** Syracuse University Library, Special Collections • **06-06** General Motors Corporation. Used with permission, GM Media Archives • **06-07** Syracuse University Library, Special Collections . • **06-08** Syracuse University Library, Special Collections . • **06-09** Syracuse University Library, Special Collections . • **06-10** Photograph courtesy of Knoll, Inc. • **06-11** Photograph courtesy of CUTCO®, Olean, New York • **06-12** Photograph courtesy of Deanne Dailey Hansen • **06-13** Syracuse University Library, Special Collections • **06-14** General Motors Corporation. Used with permission, GM Media Archives • **06-15** Photograph courtesy of Tucker Madawick, FIDSA and courtesy of Laurence Loewy of Loewy Design • **06-16** Photograph courtesy of Douglas F. Kelley, IDSA, FCSD, FRSA • **06-17** Photograph courtesy of Herman Miller, Inc. • **06-18** Photograph courtesy of the author • **06-19** Photograph courtesy of the Industrial Designers Society of America • **06-20** Photograph courtesy of the Industrial Designers Society of America • **06-21** Syracuse University Library, Special Collections • **06-22** Courtesy of the Hoover Historical Center • **06-23** From the Collection of The Henry Ford • **06-24** Photograph courtesy of the Industrial Designers Society of America. Photo by Klein's North Shore Studio. 225 Washington St., Waukegan, Illinois • **06-25** Photograph courtesy of Knoll, Inc. • **06-26** Photograph courtesy of Volkswagen of America, Inc. • **06-27** Property of AT&T Archives. Reprinted with permission of AT&T • **06-28** Courtesy of Keck-Craig, Inc. • **06-29** Photograph courtesy of Braun AG, Kronberg/Taunus • **06-30** Photograph courtesy of the Industrial Designers Society of America • **06-31** Syracuse University Library, Special Collections • **06-32** Photograph courtesy of Douglas F. Wolff, IDSA • **06-33** Photograph courtesy of Lippincott Mercer • **06-34** Photograph courtesy of Lippincott Mercer • **06-35** Photograph courtesy of Samuel M. Highberger, IDSA • **06-36** Horwitt, Nathan George. Wristwatch face. (1947). White Gold, enameled dial, 1 5/16" diam. (3.3 cm). The Museum of Modern Art, New York. Gift of the designer. Photograph © 1997 The Museum of Modern Art, New York. Digital Image © The Museum of Modern Art/Licensed by SCALA/Art Resource, NY • **06-37** Photograph courtesy of the Corning Consumer Products Company • **06-38** Photograph courtesy of the Industrial Designers Society of America • **06-39** Photograph courtesy of Raymond Spilman, FIDSA • **06-40** Photograph courtesy of Lippincott Mercer • **06-41** Photograph courtesy of Xerox Corporation • **06-42** Property of AT&T Archives. Reprinted with permission of AT&T • **06-43** Photograph courtesy of Deanne Dailey Hansen.

Chapter 7

07-01 Photograph courtesy of the Industrial Designers Society of America • **07-02** Photograph courtesy of the Industrial Designers Society of America • **07-03** Photograph courtesy of Copco • **07-04** Courtesy of Tim Cunningham, FIDSA • **07-05** Photograph courtesy of Black & Decker • **07-06** Courtesy of Thomson Consumer Electronics • **07-07** Schenectady Museum • **07-08** Photo of Johnson + Johnson *Micrin* packaging, Donald Deskey Collection, Cooper -Hewitt, National Design Museum, Smithsonian Institution • **07-09** Photograph courtesy of IBM • **07-10** Photograph courtesy of Sharon Hite • **07-11** Photograph courtesy of Lippincott Mercer • **07-12** Syracuse University Library, Special Collections • **07-13** Photograph courtesy of Richard T. Quinn • **07-14** Photograph courtesy of the author • **07-15** Syracuse University Library, Special Collections • **07-16** Syracuse University Library, Special Collections • **07-17** Courtesy of the author • **07-18** Syracuse University Library, Special Collections • **07-19** Photograph courtesy of Braun AG, Kronberg/Taunus • **07-20** Syracuse University Library, Special Collections • **07-21** Photograph courtesy of the Stanley Works, New Britain, Connecticut • **07-22** Courtesy of © Eastman Kodak Company • **07-23** Photograph courtesy of the International Housewares Association • **07-24** Syracuse University Library, Special Collections • **07-25** Courtesy of Henry Dreyfuss Associates • **07-26** Photo Time & Life Pictures, Greg Mathieson/Mai/Getty Images • **07-27** Photograph courtesy of Clair A. Samhammer, FIDSA • **07-29** Photograph courtesy of Volkswagen of America, Inc. • **07-30** Photograph courtesy of Arnold Wolf, FIDSA • **07-31** Photograph courtesy of the International Housewares Association • **07-32** Photograph courtesy of the author • **07-33** Photograph courtesy of Monte Levin, IDSA, Monte Levin Associates, Inc. • **07-34** Photograph courtesy of Herman Miller, Inc. • **07-35** Photograph courtesy of Fiskars Brands, Inc. • **07-36** Photograph courtesy of Raytheon Company and Maytag Corporation • **07-37** Photograph courtesy of Lippincott Mercer • **07-38** Courtesy of the author • **07-39** Copyright DaimlerChrysler Corporation. Used with permission • **07-40** Photograph by Earl Woods, courtesy of Herman Miller, Inc. • **07-41** Photograph courtesy of Friday Historical Business Archives, Louisville, Kentucky • **07-42** Photograph courtesy of NASA • **07-43** Property of AT&T Archives. Reprinted with permission of AT&T • **07-44** American Aviation Historical Society, photo courtesy of Jay L. Sherlock.

Chapter 8

08-01 Photograph courtesy of Lippincott Mercer • **08-02** Photograph courtesy of the author • **08-03** Photograph courtesy of Goldsmith Yamasaki Specht • **08-04** Copyright DaimlerChrysler Corporation. Used with permission • **08-05** Photograph courtesy of Lippincott Mercer • **08-06** Photograph courtesy of Wallace H. Appel, FIDSA • **08-08** Photograph courtesy of the International Housewares Association • **08-09** Photograph courtesy of Tim Cunningham, FIDSA • **08-10** Photograph courtesy of Deanne Dailey Hansen • **08-11** Photograph courtesy of Braun AG, Kronberg/Taunus • **08-12** Photograph courtesy of lenox Incorporated, Lawrenceville, NJ, USA • **08-13** Photograph courtesy of Enrico Tedeschi, Brighton, England • **08-14** Photograph courtesy of Tim Cunningham, FIDSA • **08-15** Photograph courtesy of Steelcase Inc. • **08-16** Photograph courtesy of Wallace H. Appel, FIDSA • **08-17** Photograph courtesy of American Honda Motor Co., Inc. • **08-18** Photograph courtesy of Volkswagen of America, Inc. • **08-19** Courtesy of the author • **08-20** Photograph courtesy of Tim Cunningham, FIDSA • **08-21** Photograph courtesy of Cooper C. Woodring, FIDSA • **08-22** Photograph courtesy of Cousins Design, Inc. • **08-23** Photograph courtesy of Tim Cunningham, FIDSA • **08-24** Photograph courtesy of NASA • **08-25** Photograph courtesy of Herman Miller, Inc. • **08-26** Photograph courtesy of American Honda Motor Co., Inc. • **08-27** Photo AFP Matt Campbell/Getty Images • **08-28** National Aeronautics and Space Administration (NASA), courtesy National Air and Space Museum, Smithsonian Institution (SI 84-6698) • **08-29** Photograph courtesy of Enrico Tedeschi, Brighton, England • **08-30** Photograph courtesy of Braun AG, Kronberg/Taunus • **08-31** Courtesy of the author • **08-32** Photograph courtesy of Mrs. Diana Harrison • **08-33** Photograph courtesy of Mrs. Diana Harrison • **08-34** Photograph courtesy of Cooper C. Woodring, FIDSA • **08-35** Photograph courtesy of Sony Corporation of America • **08-36** Photograph courtesy of Knoll, Inc. • **08-37** Image courtesy of Oneida, Ltd. © Onieda Ltd. All rights reserved.

Chapter 9

09-01 Photographer, E. Alan McGee, photograph courtesy of the Industrial Designers Society of America • **09-02** Photograph courtesy of the Industrial Designers Society of America • **09-03** Photograph courtesy of Herman Miller, Inc. • **09-04** Photograph courtesy of Monte Levin, IDSA, Monte Levin Associates, Inc. • **09-05** Photograph courtesy of the Industrial Designers Society of America • **09-06** Photograph courtesy of the Industrial Designers Society of America • **09-07** National Air and Space Museum, Smithsonian Institution (SI 96-15925) • **09-08** Photograph courtesy of Braun AG, Kronberg/Taunus • **09-09** Courtesy of Henry Dreyfuss Associates • **09-10** Photograph courtesy of the Industrial Designers Society of America • **09-11** Photograph courtesy of Electrolux Home Care Products North America • **09-12** Photograph courtesy of Dr. Robert Blaich, FIDSA, FRSA • **09-13** Photograph courtesy of Mrs. Lee Payne • **09-14** Photograph courtesy of the Industrial Designers Society of America • **09-15** Photograph courtesy of the Industrial Designers Society of America • **09-16** Photograph courtesy of Marc Rosen Associates • **09-17** Reproduced with Permission from Motorola Archives. © 2004, Motorola, Inc. • **09-18** Photograph courtesy of the Industrial Designers Society of America • **09-19** Photograph courtesy of Apple Computer, Inc. • **09-20** Photograph courtesy of Alex Bally, FIDSA • **09-21** Photograph courtesy of the Industrial Designers Society of America • **09-22** Photograph courtesy of the Industrial Designers Society of America • **09-23** Photograph courtesy of the Industrial Designers Society of America • **09-24** Photograph courtesy of Dr. Robert Blaich, FIDSA, FRSA • **09-25** Photograph by Claus Mroczynski courtesy of the Industrial Designers Society of America • **09-26** Photograph courtesy of the Industrial Designers Society of America • **09-27** Photograph courtesy of Kangaroo Products Company, Columbus, North Carolina • **09-28** Photograph courtesy of the Industrial Designers Society of America • **09-29** Photograph courtesy of Goldsmith Yamasaki Specht • **09-30** Courtesy of © Eastman Kodak Company • **09-31** Photograph courtesy of the Industrial Designers Society of America • **09-32** Photograph courtesy of the Industrial Designers Society of America • **09-33** Photograph courtesy of the Industrial Designers Society of America • **09-34** Photograph courtesy of the Industrial Designers Society of America • **09-35** Photograph courtesy of the Industrial Designers Society of America • **09-36** Photograph courtesy of the Industrial Designers Society of America • **09-37** Photograph courtesy of the Industrial Designers Society of America • **09-38** Reproduced with Permission from Motorola Archives. © 2004, Motorola, Inc. • **09-39** Photograph courtesy of Mazda Motor of America, Inc.

Chapter 10

10-01 Photograph courtesy of the Industrial Designers Society of America • **10-02** Photograph courtesy of the Industrial Designers Society of America • **10-03** Photograph courtesy of the Industrial Designers Society of America • **10-04** Photograph courtesy of the Industrial Designers Society of America • **10-05** Photograph courtesy of the Industrial Designers Society of America • **10-06** Photograph courtesy of Dr. Robert Blaich, FIDSA, FRSA • **10-07** Photograph courtesy of the Industrial Designers Society of America • **10-08** Photograph courtesy of the Industrial Designers Society of America, Rieder and Walsh Photography, Pittsburgh, Pennsylvania • **10-09** Photograph courtesy of Steelcase Inc. • **10-10** Photograph courtesy of the Industrial Designers Society of America, Photographer, Mark A. Steele • **10-11** Photograph courtesy of the Industrial Designers Society of America • **10-11a** Photograph courtesy of Braun AG, Kronberg/Taunus • **10-12** Photograph courtesy of the Industrial Designers Society of America • **10-13** Photograph courtesy of Sony Corporation of America • **10-14** Photograph courtesy of Sony Corporation of America • **10-15** Photograph courtesy of Sony Corporation of America • **10-16** Photograph courtesy of Tupperware • **10-17** Photograph by Don Hamerman, courtesy of the Industrial Designers Society of America • **10-18** Photograph courtesy of the Industrial Designers Society of America • **10-19** Photograph courtesy of Black & Decker • **10-20** Photograph courtesy of Herman Miller, Inc. • **10-21** Photograph courtesy of the Industrial Designers Society of America • **10-22** Photograph courtesy of the Industrial Designers Society of America • **10-23** Photograph by Jim Powell, courtesy of the Industrial Designers Society of America • **10-24** Photograph courtesy of the Industrial Designers Society of America • **10-25** Photograph courtesy of the Industrial Designers Society of America • **10-26** Photograph courtesy of the Industrial Designers Society of America • **10-27** Photograph by Jon Mazey, courtesy of the Industrial Designers Society of America • **10-28** Photograph by Jeff Noel, Northwest Images, courtesy of the Industrial Designers Society of America • **10-29** Photograph courtesy of the Industrial Designers Society of America • **10-30** Photograph courtesy of the Industrial Designers Society of America • **10-31** Photograph courtesy of the Industrial Designers Society of America • **10-32** Photograph courtesy of the Industrial Designers Society of America • **10-32a** Photograph courtesy of Niels Diffrient, FIDSA • **10-33** Photograph courtesy of the Industrial Designers Society of America • **10-34** Photograph courtesy of the Industrial Designers Society of America • **10-35** Photograph courtesy of the Industrial Designers Society of America • **10-36** Photograph courtesy of Braun AG, Kronberg/Taunus

Chapter 11

APPEL Photograph courtesy of Wallace Appel, FIDSA • **BACH** Photograph courtesy of Anita S. Bach • **BECVAR** Photograph courtesy of the Industrial Designers Society of America • **BEHRENS** Deutsches Technikmuseum Berlin, AEG-Archiv, © 2004 Artists Rights Society (ARS), New York/VG Bild-Kunst, Bonn • **BEL GEDDES** Photo courtesy of the estate of Edith Lutyens Bel Geddes, Norman Bel Geddes Collection, Harry

Ransom Humanities Research Center • **BLAICH** Photograph courtesy of Dr. Robert Blaich, FIDSA, FRSA • **BORDINAT** Photograph courtesy of the Industrial Designers Society of America • **CHAPMAN** Portrait of David Chapman, David Painter Collection. George Nelidoff, Chicago, photographer. Courtesy of the Art Institute of Chicago • **DAILEY** Photographer Moffett Studio, photograph courtesy of the Industrial Designers Society of America and Deanne Dailey Hansen • **DESKEY** Portrait of Donald Deskey, Donald Deskey Collection, Cooper-Hewitt, National Design Museum, Smithsonian Institution • **DIAMOND** Freda Diamond Collection, Archives Center, National Museum of American History, Behring Center, Smithsonian Institution • **DIFFRIENT** Photograph courtesy of the Industrial Designers Society of America and permission of Niels Diffrient, FIDSA • **DOBLIN** Permission to use image granted by Doblin, Inc. • **DOHNER** Permission to use image granted by Lippincott Mercer. • **DREYFUSS** Photograph courtesy of Henry Dreyfuss Associates • **ENGEL** Syracuse University Library, Special Collections • **EXNER** Syracuse University Library, Special Collections • **FERAR** Image courtesy of Curt Bailey, President, Sundberg-Ferar, Inc. • **FULLER** Photograph courtesy of the Industrial Designers Society of America • **FULTON** Image courtesy of the author with permission of Priscilla Lambert Fulton • **GLASS** Photograph courtesy of Victoria Matranga • **GRASER** Photograph courtesy of Gladys Graser • **HAGGSTROM** Photograph courtesy of the Industrial Designers Society of America • **HAUSER, J.** Photograph courtesy of the Industrial Designers Society of America • **HIGHBERGER** Photograph courtesy of Samuel M. Highberger, IDSA • **HUBBARD** From the Collection of Roycroft Arts Museum, Bruce Lydell, East Aurora, New York, 1916 catalog p. 2. Photograph by the author • **KAUFFMAN** Photograph by Kenneth Love, courtesy of the Western Pennsylvania Conservacy • **KNOLL, F.** Photograph courtesy of Knoll, Inc. • **11-25** Image courtesy of the author • **LATHAM** Photographer, Austin Field, Chicago, photograph courtesy of the Industrial Designers Society of America • **LIPPINCOTT** Photograph courtesy of Lippincott Mercer • **LOEWY** Use of photograph courtesy of Laurence Loewy of Loewy Design • **MADAWICK** Photograph courtesy of Tucker Madawick, FIDSA • **MAYHEW** Photograph courtesy of Fritz Mayhew • **MIES VAN DER ROHE** Ludwig Mies van der Rohe. (c.1956). Photograph courtesy, The Museum of Modern Art, New York. Digital Image © The Museum of Modern Art/Licensed by SCALA/Art Resource, NY, • **MITCHELL** Syracuse University Library, Special Collections • **MULLER-MUNK** Permission to use image granted by Peter Muller Munk Associates • **MYERS** Photograph courtesy of the Industrial Designers Society of America • **NELSON** Photograph courtesy of the Industrial Designers Society of America • **NOYES** Syracuse University Library, Special Collections • **OROS** Syracuse University Library, Special Collections • **OTTO** Syracuse University Library, Special Collections • **RAMS** Image courtesy of Braun GmbH • **REDDIG** Photograph courtesy of William E. Reddig • **REINECKE** Photograph courtesy of Barbra Reinecke Sedory • **ROHDE** Photograph courtesy of Herman Miller, Inc. • **SAARINEN, EERO** Photograph courtesy of Knoll, Inc. • **SAARINEN, ELIEL** Eliel Saarinen, 1873-1950 (Designer) *Cranbrook School for Boys, Dining Hall Headmaster's Armchair*, 1928 , Oak with cast-bronze medallion, Collection of Cranbrook Art Museum, Bloomfield Hills, Michigan. Gift of Cranbrook School (CAM 1979.22) Photography (c) Cranbrook Art Museum • **SCHERR** Image courtesy of Mary Ann Scherr • **SCHRECKENGOST** Photograph by Herb Ascherman courtesy of Viktor Schreckengost Foundation • **SINEL** Image courtesy of the author • **SMITH, D.** Photograph courtesy of David B. Smith, IDSA • **SPECHT** Photograph courtesy of Paul B. Specht, FIDSA • **STEINHILBER** Photograph courtesy of Budd Steinhilber, FIDSA • **STEVENS** Photograph by Carroll Studios. Inc., courtesy of the Milwaukee Institute of Art & Design • **SUNDBERG** Image courtesy of Curt Bailey, President, Sundberg-Ferar, Inc • **TEAGUE** Photo courtesy of the Industrial Designers Society of America. © Walter Dorwin Teague Associates, Inc • **VASSOS** Photograph courtesy of the Industrial Designers Society of America • **VIEMEISTER, R.** Photograph courtesy of Tucker Viemeister, FIDSA• **WALKER** Syracuse University Library, Special Collections • **WRIGHT, F. L.** Frank Lloyd Wright (b. 1867-d. 1959), c. 1890-1895, Photographer: unknown. Collection of The Frank Lloyd Wright Preservation Trust, H & S H177 • **WRIGHT, R.** Photograph courtesy of the Special Collections Research Center, Syracuse University Library; with permission of the Russel Wright Center, Manitoga, New York.

All other biographical images are courtesy of the author.

Bibliography

Books

Abercrombie, Stanley, *George Nelson:The Design of Modern Design*, MIT Press, 1994.

Adamson, Glenn, *Industrial Strength Design*, MIT Press, 2003

Amato, Ivan, *Stuff,* Basic Books, HarperCollins, 1997

Ash, Russell, *The Top Ten of Everything*, DK Publishing, Inc., 1996.

Baker, Eric, *Great Inventions-Good Intentions,* Chronicle Books, 1990

Bamford, James, *The Puzzle Palace*, Houghton Miflin Company, 1982

Banham, Rayner, *Theory and Design in the First Machine Age,* MIT Press, 1992, reprint of 1st edition, Architectural Press, London, 1960.

Basalla, George , *The Evolution of Technology,* Cambridge University Press, 1988.

Bayley, Stephen , *In Good Shape*, Van Nostrand Reinhold Company, 1979.

Bel Geddes, Norman, *Horizons,* Little Brown & Company, 1932

Birren, Faber, *Color and Human Response,* Van Nostrand Reinhold Co., 1978

Birren, Faber, *Color,* Citadel Press, 1963.

Blaich, Robert with Janet, *Product Design and Corporate Strategy*, McGraw-Hill, Inc., 1993

Boorstin, Daniel J., *The Creators,* Random House, 1992.

Boorstin, Daniel J., *The Image,* Vintage Books, 1992; Originally published 1961.

Bourke, Bob, *Designing for Studebaker,* 1984

Bridges, John, *Bob Bourke Designs for Studebaker* (Draft)

Brown, Travis, *Historical First Patents,* Scarecrow Press, Inc., 1994

Bryan, C.D.B, The National Air and Space Museum, Vol. 1, *Air, The Story of Flight,* A Bantam Book published by Harry N. Abrams Inc., 1979.

Bryant, John H., and Cones, Harold N., *The Zenith Trans-Oceanic,* Schiffer Collector Books, 1995

Bunch, Bryan, and Alexander Hellemans, *The Timetables of Technology,* edited by Touchstone, Simon & Schuster, 1993

Burke, James, *The Day the Universe Changed,* Little, Brown & Co., 1985.

Burke, James, The Pinball Effect, Back Bay Books, Little, Brown & Co., 1996

Burke, James, *Connections*, Little, Brown & Co., 1978.

Byars, Mel, with Arlette Barré Despond, *100Designs/100 Years-Innovative Designs of the 20th Century*, RotoVision SA, Switzerland, 1999

Caney, Steven, *Invention Book*, Workman Publishing Company, Inc., 1985

Caplan, Ralph, *By Design*, St. Martin's Press, 1982.

Casey, Louis S., *Curtiss-The Hammondsport Era 1907-1915,* Crown Publishers, 1981.

Chajet, Clive, and Tom Shactman, *Image by Design,* AddisonWesley Publishing Company, 1991.

Chapelle, Howard I., *The History of American Sailing Ships*, W.W. Norton & Company, Inc., N.Y., 1935

Cheney, Margaret, *Tesla-Man Out of Time,* Barnes & Noble, 1993

Conaway, James, *The Smithsonian,* Alfred A. Knopf, Inc., and Smithsonian Books, 1995.

Cowan, Ruth Schwartz, *More Work for Mother,* Basic Books, Inc. 1983.

de Noblet, Jocelyn, *Industrial Design, Reflection of a Century*, published by Imprimerie Clerc SA, Saint-Armand-Montrond, 1993, in connection with exhibit of same name at the Grand Palais, Paris.

DiNoto, Andrea, *Art Plastic* designed for Living, Abbeville Press, 1984

Doblin, Jay, *One Hundred Great Product Designs,* Van Nostrand Reinhold, 1970

Dormer, Peter, *Design Since 1945*, Thames and Hudson, Ltd., 1993.

Edsall, Larry, *Concept Cars*, Barnes & Noble, Inc., 2003

Edwards, Sandra, and ID Magazine, Product Design 2, PBC International, 1987

Egan, Phil, *Design and Destiny the Making of the Tucker Automobile.,* On the Mark Publications, 1989

Fawdry, Kenneth, and Marguerite, *English Dolls & Toys,* Ernest Benn Ltd. 1979.

Ferebee, Ann, *A History of Design from the Victorian Era to the Present*, Van Nostrand Reinhold Company, 1970.

Ferguson, Eugene S., *Engineering and the Mind's Eye*, MIT Press, 1992

Flatow, Ira , *They All Laughed,* Harper Perennial, 1992.

Fuller, R. Buckminster, *Operating Manual for Spaceship Earth,* Southern Illinois University Press, 1969; Republished by Aonian Press Inc. 1976.

Furnas, C.C., *The Next Hundred Years.* Reynal & Hitchcock, 1936.

Gartman, David, *Auto Opium,* Routledge, 1994.

Gelernter, David, *1939-The Lost World of the Fair,* The Free Press, Simon & Schuster, Inc.,1995

Gelerntner, David, *Machine Beauty,* Basic Books, 1998

Gilbert, Ann, *'40s/'50s Designs & Memorabilia,* Avon Books, 1993.

Gilbert, Gorman, and Samuels, Robert E., *The Taxicab, An Urban Transportation Survivor.*

GM Styling Staff, *Styling: The Look of Things* 1955.

Gorb, Peter, *Design Talks,* The Design Council, 1988

Grove, Eric, *World War II Tanks,* Longmeadow Press, 1987

Halberstam, David ,*The Fifties*, Ballentine Books, 1993.

Hauffe, Thomas, *Design, an Illustrated Historical Overview* Barron's Educational Series, Inc., 1996; First published 1995 in Germany by DuMont Buchverlag GmbH und Co. Kommanditgesellschaft, Köln.

Heath, Adrian;,Heath, Ditte, and Jensen, Aage Lund; *300 Years of Industrial Design.* Watson-Guptill, 2000

Heskett, John, *Industrial Design,* Thames & Hudson, 1985.

Heskett, John, *German Design 1870-1918*, Taplinger Publishing Co., Inc., 1986

Hiesinger, Kathryn B., and George H. Marcus, *Landmarks of Twentieth Century Design,* Abbeville Press 1993.

Hines, Thomas S., *Irving Gill and the Architecture of Reform*, Monacelli Press, 2000

Hollis, Richard, *Graphic Design-A Concise History,*Thames and Hudson Ltd., 1994

Hughes, Thomas P., *American Genesis*, Penguin Books, 1989

Huxford, Sharon, and Bob, *Fiesta,* Collector Books, 1992

IDSA, Industrial Design in America, Farrar, Straus & Young, 1954

Ikenson, Ben, *Patents-Ingenious Inventions*, B;lack Dog & Leventhal Publishers, Inc. 2004

Jonas, Susan, and Marilyn Nissenson, *Going, Going, Gone: Vanishing Americana,* Chronicle Books, 1995.

Julier, Guy, *Encyclopaedia of 20th Century Design and Designers*, Thames and Hudson, Ltd., 1993.

Keegan, John, *A History of Warfare,* Vintage Books, 1994

Kerr, Ann, *Russel Wright Designs,* Collector Books, 1990

Ketchum, Richard M., *The Borrowed Years,*Random House, 1989.

Kovel, Ralph, and Terry. *Know your Collectibles*, Crown Publishers, Inc., 1981

Lifshey, Earl, *The Housewares Story,* the National Housewares Manufacturers Association (NHMA), 1973.

Loewy, Raymond , *Never Leave Well Enough Alone*, Simon and Schuster, 1951.

Loewy, Raymond , *Industrial Design*, Overlook Press, 1979.

Lorenz, Christopher , *The Design Dimension*, Basil Blackwell Ltd., 1986.

Matranga, Victoria, *America at Home,* National Housewares Manufacturers Association, 1997

McDermott, Catherine, *Twentieth Century Design*, Overlook Press, 2000

Meikle, Jeffrey L., *American Plastic,* Rutgers University Press, 1995.

Messadié, Gerald *Great Modern Inventions,* W & R Chambers Ltd., 1991.

Miller, Donald L., *City of the Century,* Touchstone, 1996

Mingo, Jack, *How the Cadillac Got It's Fins,* Harper Collins Publishers, 1994.

Moholy-Nagy, Laslo , *Vision in Motion*, Paul Theobald, Chicago, 1947.

Nelson, Walter Henry, *Small Wonder-The Amazing Story of the Volkswagen,* Little, Brown & Company, 1965, 1967,1970.

Newbury, Stephen, *The Car Design Yearbook*, Merrell Publishers Limited, 2003
Norman, Donald A., *Turn Signals are the Facial Expressions of Automobiles*, Addison-Wesley Publishing Co., 1992
Panoti, Charles, *Panati's Extraordinary Origins of Everyday Things*, Perennial Library, Harper & Row, 1987
Patton, Phil , *Made in U.S.A.*, Penguin Books,1992.
Petroski, Henry, *Invention by Design*, Harvard University Press, 1996.
Petroski, Henry, *The Evolution of Useful Things*, Alfred A. Knopf, 1992.
Petroski, Henry, *To Engineer is Human,* Vintage Books, a division of Random House, 1992.
Petroski, Henry, *Remaking the World*, Vintage Books, a division of Random House, 1999.
Petroski, Henry, *Small Things Considered*, Alfred A. Knopf, 2003
Philip. Craig, World's Great Small Arms, Brown packaging Books, Ltd., 1993
Pile, John, *The Dictionary of 20th Century Design,* Da Capo Press1994 (orig publ. 1990).
Platt, Richard, *Smithsonian Visual Timeline of Inventions,* Dorling Kindersley, Ltd., London, 1994.
Preble, Duane, *Man Creates Art Creates Man,* Canfield Press, 1973.
Pulos, Arthur J., *American Design Ethic*, MIT Press, 1983.
Pulos, Arthur J., *The American Design Adventure*, MIT Press, 1988.
Pursell, Carroll W., *The Machine in America,* Johns Hopkins University Press, 1995.
Rams, Dieter, Design, *Dieter Rams &,* Gerhardt Verlag, 1980 and 1981
Readers Digest Association, *Stories Behind Everyday Things*, 1980
Rybczynski, Witold, *Home,* Penguin Books, 1986
Rybczynski, Witold , *Looking Around*, Penguin Books, 1992.
Rybczynski, Witold, *City Life,* Scribner, 1995
Rybczynski, Witold, *One Good Turn-A Natural History of the Screwdriver and the Screw*, Scribner, 2000.
Schönberger, Angela; *Raymond Loewy:Pioneer of American Industrial Design,* Prestel, Germany 1991.
Schutte, Thomas F., The Tiffany-Wharton Lectures, *The Art of Design*, University of Pennsylvania Press, 1975
Secrest, Meryle , *Frank Lloyd Wright*, Harper Perennial, 1993.
Shirer, William L., *The Rise and Fall of the Third Reich,* Simon and Schuster, 1959
Staten, Vince, *Did Monkeys Invent the Monkey Wrench?,* Touchstone, Simon & Schuster, 1997
Stern, Jane, and Michael, *Encyclopedia of Pop Culture,* Harper perennial, 1992.
Tambini, Michael, *The Look of the Century,* DK Publishing, Inc., 1996
Tenner, Edward. *Our Own Devices*, Alfred A. Knopf, 2003
Toledo Museum of Art, *The Alliance of Art and Industry—Toledo Designs for a Modern America*, Hudson Hills Press, NY, 2002.
Trager, James, *The People's Chronology,* Henry Holt and Company, 1994
Tryckare, Tre, *The Lore of Flight,* Cagner & Company, Sweden, 1970.
Van Dulken, Stephen, Inventing the 20th century, New York University Press, 2000
Wallechinsky, David, and Wallace, Irving, *The People's Almanac*, Doubleday and Company, Inc., 1975
Wallechinsky, David, and Wallace, Irving, *The People's Almanac*, #2, Bantom Books, 1978.
Weale, Elke C., John A. Weale, and Richard F. Barker, *Combat Aircraft of World War II*, Arms and Armour Press,1977.
Wilkes: Encyclopedia of Architecture, John Wiley & Sons, Inc. 1989.
Wingate, P.J. , *The Colorful DuPont Company*, Serendipity Press, 1982.
Wingler, Hans, *The Bauhaus*, MIT Press, 1969.
Whitford, Frank, *Bauhaus,* Thames and Hudson Ltd., London, 1984.
Zunz, Olivier, *Why the American Century*, University of Chicago Press ,1999-

Periodicals

Air & Space- Aug/Sept 1996 Vol 11, No. 3, *Who Made the Mustang?*, Peter Garrison; **American Aviation Historical Journal**, Vol 42, No. 4; **American Society of Industrial Designers, Newsletters** 1962-1964; **Appliance Magazine** June 1994 *The Evolution of Convenience,* Michel Sanders; **Automotive Engineering**- January, 1996 *The first 100 Years of Transportation Safety: Part 1*, by Anthony J. Yanik ; **Automobile Quarterly**- Vol 3 No. 4, 1965, *Alexis de Sakhnoffsky*, Beverly Rae Kimes • Summer, 1971 *The New Nostalgia*, by Richard M. Langworth • Vol 10 No.3, 1972. *Studebaker,* Maurice D. Hendry • Vol 16-No.1 1978, *Hupmobile: Always a Good Car,* Jeffrey I. Godshall • Vol 16 No 4, 1978 *The All-Steel World of Edward Budd,* Stan Grayson, • Vol 18 No. 8 *The Slippery Shapes of Paul Jaray*, by Jerry Sloninger • Vol 26 No 93, 1988 *Wielding Thunderbolts*, and, *Tremulis; The Genius Behind the Tucker,* by Phil Egan ; **Better Homes & Gardens**, Sept 1929; **Carnegie Magazine**- May/Jun 1987 *Robert Lepper and the Machine Age,* by Richard Guy Wilson ; **Chicago Tribune**, Nov 21, 1993 *The Classics*; **Classic Car,The**- March 1994 *Classic Marmons & Classic Memories*, Dorwin Teague; **Collectible Automobile**- June 1995. *Homer LaGassey,* Tim Howley,; **Continental Comments**- March/April 1997, *The Bob Gregorie Story, Part I*, Tucker Madawick • May/June 1997, *The Bob Gregorie Story, Part II*, Tucker Madawick • July/Aug 1997, *The Bob Gregorie Story, Part III*, Tucker Madawick • Sept/Oct 1997, *Clay Modeling: The Designers Artform,* Tucker Madawick; **Cooper Hewitt Magazine-**1993-1997; **Current Commentary**, Hall of History; **Designscape**, New Zealand ID Council, February, 1977 article on *Jo Sinel* , Gifford Jackson; **Design Trends**, Nowland & Schladermundt; .**Fortune**- February, 1934, *Both Fish and Fowl* • February 1959, *Old Business, This Industrial Design,* by Seymour Freedgood ; **Gristmill, The** - June 2002, *L. Garth Huxtable, Industrial Designer for Millers Falls*, Randy Roeder; **Hennepin History-** Summer 1995 *Betty Crocker's Return from World War II*, Bruce Knight ; **Industrial Designer's Institute Newsletters** 1962-1964; **ID Magazine**- Feb 1954 *The Studebaker Story*, John Wheelock Freeman • Apr 1958 *Art Nouveau and All That,* Edgar Kaufmann, Jr. • May 1958, *Magnetic Tape in Action,* Arthur Greyor • Jun 1966 *Consultant Design: A Retrospective View,* Arthur J. Pulos • Jun 1971 *A Half Century of Design* • Jul/Aug 1981 *Coldspots and Heaters*, Jeffrey L. Meikle • Mar/Apr 1982, MoMA *Good Design Programs Changed Historic US Taste,* Richard Horn • May/Jun 1982, *Fashion of the '50s*, Penny Sparke • Mar/Apr 1983, *Just a Phone Call Away,* Michael Sorkin • May/Jun 1983, *The Dream Machine Comes of Age,* Kenneth Lalen • Jul/Aug 1983 *The Bicycle,* Iain Boal • Jul/Aug 1983 *Eva Zeisel-Humanistic Design for Mass Production*, Sarah Bodine • Nov/Dec 1983 *From Aspen to Memphis*, Jeffrey L. Meikle • Nov/Dec 1983 *Tastemaking in America: the birth of mass design*, Karen Davies • Jan/Feb 1984, *Made in Japan.* John Heskett • May/Jun 1985 *Singer Design 1850-1985,* Arthur Pulos • Sept/Oct 1985 *Switching to Swatch,* Annetta Hanna • Jan/Feb 1986 *Alan Heller,* Annetta Hanna • Jan/Feb 1986 *Vacuum Streamliners,* Jeff Brewer • Mar/Apr 1986, *Homage to George Nelson,* Randolph McAusland • Nov/Dec 1986 *Loewy,* Jeff Meikle • Nov/Dec 1986 *The Designer's Car*, David Schultz • Jan/Feb 1988 *Designing for a Giant,* Randolph McAusland • Jan/Feb 1988 *Joseph Urban,* Timothy Rub • Jan/Feb 1989 *The Next Contender,* Chee Pearlman • Jan/Feb 1989 *Munich Art Nouveau,* Nick Backlund • Nov/Dec 1989 *Super Trains,* Hugh Aldersey-Williams • Nov/Dec 1989, *Blazed Trails,* Douglas A. York, Jr; • Jan/Feb 1990, *Silicon Valley Stories,* Chee Pearlman • Jan/Feb 1991 *Chicago*, Chee Pearlman • Jan/Feb 1991 *Industrial Icons,* Christian K. Laine • Nov/Dec 1991 *Satellite Studios*, Chee Pearlman • Jan/Feb 1992, *Noguchi's Bakelite Nanny,* Andrew Olds • Jan/Feb 1992, *Superbikes,* Phil Patton • Jan/Feb 1992 *George Nelson,* Ralph Caplan • Mar/Apr 1992, *Teaching an Old Dog New Tricks*, John Heskett • Mar/Apr 1992 *Material World,* Frederique Krupa • Sept/Oct 1992 *Loewy's icon of Streamlined Design,* Alexandra Anderson-Spivy • Mar/Apr 1993 *The Lockheed Constellation*, Phil Patton • Sept/Oct 1993 *The Telephone,* Ellen Lupton • Sept/Oct 1993 *Getting Back on Track,* Phil Patton • Mar/Apr

1994 *A Secret History of Design in LA,* Michael McDonough and Gere Kavanaugh • Mar/Apr 1994 *Nike comes Down to Earth*, Phil Patton • May/Jun 1994 *Mr. Sunbeam,* John Heskett • May/Jun 1994 *Hard Hats,* Phil Patton • Jan/Feb 1995 *Merchandising Modernity,* Michael Kaplan • Nov 1995 *The Business of Color,* Phil Patton • Jan/Feb 1996 *Designing History,* Michael Rock • Mar/Apr 1996 *Lilly Reich Remembered,* Miguel Baltierra • Mar/Apr 1997, *Henry Dreyfuss, Designing the American Dream,* by Russel Flinchum ; **Industrial Designers Society of America Newsletters**, 1965-2000; **Innovation**, Industrial Designers Society of America Journal, Fall 1986 *The DC-3:The Design of a Classic*, Michael J. Szablak • Spring 1991 *The Patents of a Design Pioneer,* Nancy Perkins • Winter 1995 *Brooks Stevens*, Thomas David • Spring/Summer 1995 *Century of Cars,* Carroll Gantz ; **Invention & Technology-** Summer 1987 *The First Plastic*, Robert Friedel • Fall, 1989, *The Path of Least Resistance,* James J. Flink • Spring/Summer 1990 *A Most Invented Invention,* David B. Sicilia • Spring 1992 *The Bicycle Story,* David V. Herlihy • Fall1994 *The Man Who Invented the PC,* Lamont Wood • Spring 1996 *Troubled Marriage-Raymond Loewy and the Pennsylvania Railroad,* Glenn Porter • Summer 1996 *What Makes Bell Labs Great,* T.A. Heppenheimer • Summer 1996, *Unplanned Obsolescence,* James J. Flink • Fall 1996 *Kettering*, Oliver E. Allen • Fall 1997 *The Need to Read Speed,* Michael Lamm • Winter 1997 *Gas Pumps*, Michael Karl Witzel • Spring 1998 *The Toys That Built America*, Henry Petroski • Fall 1998, *The Earl of Detroit*, Michael Lamm • Winter 1998, *The Instant Building*, Michael Lamm • Fall 1999 *The Woman Who Invented the Dish-Washing Machine*, J.M. Fenster • Winter 1999, *Howard Hughes*, by T.A. Heppenheimer • Spring 2000 *Radio Hits the Road*, Michael Lamm; **Iron Age**-Special Bicentennial Edition, April 12, 1976 *Men and Machines* ; **Leading the way**, 1994 Thompson Consumer Electronics; **Life**, Collectors Edition, Feb 1996 *Hot for the Road; 100 years of the automobile in America.*; **Link**-the Cleveland Institute of Art magazine, Summer 1987 *The Art of Car Design,* Hugh Greenlee; **Metropolis**- March 1997, *The Man Inside the Brown Suit*, Christine Liotta; **Newsweek**-Extra: Winter 1997-1998 *The Power of Invention,*; **New Yorker**- April 13, 1987, Profiles, *The Present Moment,* Suzannah Lessard,; **New York Times Magazine-** Home Design, Apr 5, 1998, *Feat of Clay,* Penelope Rowlands, and *Material Man,* by Akiko Busch • Dec 13, 1998, *The Shock of the Familiar* • Apr 18, 1999, *The Best Ideas, Stories and Inventions of the Last 1000 Years* • Nov 30, 2003, *Inspiration-Where Does It Come From?*; **Philco World**, Vol. 7 1954; **Railroad Heritage Magazine**- Spring 2000, *Diesel Styles and Styling*, Greg Palumbo and John Gruber; **Scientific American-** Aug 1977, *The History of the Airflow Car,* Howard S. Irwin • Nov. 1990, *Science, Technology, and the Western Miracle,* Nathan Rosenberg and L.E. Birdsell • Sept 1994, *The Aluminum Beverage Can,* William F. Hosford and John L. Duncan; **Smithsonian**-Dec.1991 *Sleek but Square,* Phil Patton • December 1997 *This is Not Your Father's Automobile,* Bruce Watson • Oct 2000, *Art Nouveau*, Stanley Meisler ; **Special Interest Autos**- Jul-Aug 1974 *The First Mercury,* Michael Lamm and David L. Lewis • June 1991 *1936 Stout Scarab,* Rich Taylor • March-Apr 1996 *The Coachbuilders, Brewster,* Michael Lamm • May/June 1996 *Amos Northrop and Le Baron* , Michael Lamm • July/Aug 1996 *Elwood Engel,* Michael Lamm; **Time Magazine**, Special Issue "American Visions", Spring 1997; **Turning Wheels**, Studebaker Drivers Club, Inc. June 1996, *1959 Larks*, Fred K. Fox

Other References

American Society of Industrial Design, Annual Reports; Comptons Pictured Encyclopedia, 1944; Consolidated Vultee Aircraft Corporation documents; Cuisinart, History: Cuisinarts, Susan Dumont-Bengston; Encyclopedia of Collectables, Time-Life Books, 1978; DeLorean, The History of the DeLorean Motor Company, James Espy, Spam Luncheon Letter, June 1996; Dixie Cup Company History, by Anke Voss-Hubbard, 1995; Dupont, The Autobiography of American Enterprise; General Electric, *A Walk Through the Park*, Elfin Historical Society; *The General Electric Story*, 1876-1986, Photo- History, Hall of History, 1989; Hall of History, Schenectady, NY; General Motors, *Styling: The Look of Things, GM Styling Staff, 1955;* ID Magazine, Annual Design Reviews; Industrial Designers Institute (IDI) Annual Design Awards, 1951-1965; Industrial Designers Society of America (IDSA), Industrial Design Excellence Awards, 1980-2000; IDSA Directories, 1965-2000; Information Please Almanac, 1993; Lightolier, Journey © 1994; NCR, 100 Anniversary History; Pratt Industrial Design Department Timeline, 1996; Industrial Design at Pratt Institute, 1936; Pratt History, 1974; RCA, *The Sound Connection*, History Highlights; Westinghouse Design Manuals; World Almanac, St. Martin's Press, 1998; World Almanac, New York Times, 1987; Xerox Fact Book, 1997

Index